JOSEPH T. BOHANNON

Fire Protection

for the Design Professional

Contributors

Wayne E. Ault, P.E.
  Manager of Engineering
  "Automatic Sprinkler Corporation of America
  Cleveland, Ohio

Charles F. Averill
  Manager, Special Hazards Department
  Grinnell Company, Inc.
  Providence, Rhode Island

Jack A. Bono, P.E.
  Assistant Chief Engineer—Fire Protection
  Underwriters Laboratories, Inc.
  Northbrook, Illinois

William H. Doyle, P.E.
  Chief Chemical Engineer
  Factory Insurance Association
  Hartford, Connecticut

Charles L. Ford
  Project Leader, Fire Extinguishants
  Organic Chemicals
  E. I. duPont de Nemours & Co.
  Wilmington, Delaware

Arthur B. Guise, P.E.
  Consulting Engineer
  Rolf Jensen & Associates, Inc.
  Marinette, Wisconsin

Walter M. Haessler, P.E.
  Florida State Fire College
  Ocala, Florida

Robert D. Harger
  Harger Lightning Protection, Inc.
  Libertyville, Illinois

Boyd A. Hartley, P.E.
  Associate Professor
  Department of Fire Protection &
    Safety Engineering
  Illinois Institute of Technology
  Chicago, Illinois

C. F. Hedlund, P.E.
  Chief Electrical Engineer
  Factory Mutual Engineering Association
  Norwood, Massachusetts

Rolf Jensen, P.E.
  Professor & Chairman
  Department of Fire Protection &
    Safety Engineering
  Illinois Institute of Technology
  Chicago, Illinois

President
  Rolf Jensen & Associates, Inc.
  Deerfield, Illinois

Gerald L. Maatman, P.E.
  President
  National Loss Control Service Corporation
  Long Grove, Illinois

D. N. Meldrum
  President
  National Foam System, Inc.
  West Chester, Pennsylvania

A. J. Mercurio
  Executive Engineer
  Factory Insurance Association
  Hartford, Connecticut

Charles B. Miller, P.E.
  General Manager, International Operations
  "Automatic" Sprinkler Corporation of America
  Cleveland, Ohio

James W. Nolan
  President
  James W. Nolan Company
  Chicago, Illinois

Marshall E. Petersen, P.E.
  Consulting Engineer
  Rolf Jensen & Associates, Inc.
  Deerfield, Illinois

Patrick E. Phillips, P.E.
  Senior Fire Protection Engineer
  U.S. Atomic Energy Commission
  Las Vegas, Nevada

William Pritsky, P.E.
  Code Engineer
  Building Products
  The Aluminum Association
  New York, New York

J. R. Williams, P.E.
  Staff Consultant
  National Foam System, Inc.
  West Chester, Pennsylvania

Herbert Witte, P.E.
  Consulting Engineer
  Underwriters Laboratories, Inc.
  Northbrook, Illinois

Fire Protection
for the Design Professional

edited by **Rolf Jensen, P.E.**

Professor and Chairman
Department of Fire Protection and Safety Engineering
Illinois Institute of Technology
and
President
Rolf Jensen and Associates, Inc.
Fire Protection Engineers & Code Consultants

Sponsoring Editor: Don DeMichael, Editor and Associate Publisher, *Actual Specifying Engineer*

CAHNERS BOOKS
A Division of Cahners Publishing Company, Inc.
89 Franklin Street, Boston, Massachusetts 02110

Second Printing

Library of Congress Cataloging in Publication Data
Main entry under title:

Fire protection for the design professional

The contributions originally appeared in Actual
specifying engineer from 1967 to 1973.
Includes index.
1. Fire prevention.   I.   Jensen, Rolf, 1929–
TH9145.F53   628.9'22   75–9508
ISBN 0–8436–0152–3

Library of Congress Catalog Card Number: 75–9508

ISBN: 0–8436–0152–3

Design and composition by Jay's Publishers Services.
Printed in U.S.A. by Halliday Lithograph Corporation.

Contents

Foreword

"The three greatest causes of fire are men, women and children." So said Percy Bugbee, for nearly 50 years General Manager of the National Fire Protection Association. While we cannot remove people from the face of the earth, we can, through fire-safe design and engineering of the built environment, contribute substantially to reducing the annual toll of lost life and destroyed property from fire.

This book, edited by a recognized and respected fire protection engineer with contributions from twenty leading fire protection engineers, experts in specific areas, will provide the design professional with valuable insight and guidance to fire protection of the built environment.

In this volume, set down probably for the first time in one handy source, is reference material on building codes, their application and misapplication; information about regulatory authorities; testing laboratory approval methods; economics of fire protection—material that has previously appeared only piecemeal.

Gathered together is information on water supply and water extinguishing systems taking the reader from the source of supply through the building to the sprinkler head or nozzle tip. Similarly, special extinguishing agent systems for special hazards are discussed, as are fire alarm systems and detection systems.

We live today in the era of the consumer. Where once the entrepreneur, the businessman large or small, held the reins of power (to be succeeded by the forces of organized labor during the 1930s, '40s, and '50s), the consumer now sits in the saddle. He requires "quality" in life not just "quantity"—his desires are voiced for clean air, clean water, nonflammable clothing, healthful working conditions, a fire-safe environment. Use of this book can provide the knowledge applicable to meeting the fire safety needs of consumers—who are, after all, people—the men, women and children whom we serve and protect.

The Society of Fire Protection Engineers extends its sincere thanks to Rolf Jensen, the editor, for his editorial skills, his tenacity, his patience in seeing this often-arduous project through to completion—to the twenty contributing authors of the original *Actual Specifying Engineer* magazine articles, now compiled in this one source, for sharing their expertise—to Joe Johnson, SFPE President at the inception of the project, and his Executive Committee for their foresight in authorizing Society sponsorship of this activity—and to the members of the Society of Fire Protection Engineers for their day-to-day, week-to-week and year-to-year professional efforts to make our environment safe from fire.

D. PETER LUND, CAE
Executive Director
Society of Fire Protection Engineers
Boston, Massachusetts

Preface

The technology of fire protection engineering has been emerging rapidly in the last decade—so fast that it is almost presumptuous to write a book on the subject and believe that its contents will reflect contemporary knowledge. Yet the need is great to put down what we know as best we can.

To most designers, fire protection in a building project means compliance with codes and standards and the associated rulings of enforcing authorities. But codes and standards are limited by the consensus nature of our standards-making system in the United States. They all have weaknesses. Too often they represent a dated version of the state of the fire protection art when written—or worse yet, a reaction to a recent disaster or the best compromise a committee could reach.

Yet this system, despite its weaknesses, is our best effort yet, and to the design professional with limited expertise in fire protection, it represents his only reference source.

This book is directed to design professionals who must do business in the everyday, real world largely following these codes and standards. Hopefully, it will provide the reader with a better understanding of fire protection and of codes and standards. To some, it will provide a basis for better and more economical fire protection design.

It represents a seven-year writing effort. It started in 1967 with the hope that the result would be a fire protection engineering textbook. It ended as a compilation of articles about significant contemporary fire protection engineering problems, all of which have been published in *Actual Specifying Engineer* Magazine.

This book is organized into four parts intended to relate to usually definable areas of usage:

Part I (Chapters 1 to 6) provides basic background information to the U.S. codes and standards system, the responsibility of the designer in fire protection, the methods and duties of approving authorities, and the approval activities of Underwriters Laboratories, Inc. It also provides information on fire protection economics and the structural aspects of fire protection engineering. All this is generally needed background information useful to all design professionals. Some of the material on codes, standards, and approvals systems may seem inappropriate for everyday use, but it is the system in the United States, and to perform in it, the design engineer must know it.

Part II (Chapters 7 to 13) deals with fire protection systems that use water or water with additives. This section is undoubtedly the most important part of the book for the mechanical, electrical, or plumbing engineer. Its engineering technology is based on classic hydraulics and fluid mechanics.

Part III (Chapters 14 to 19) discusses extinguishment systems employing agents other than water. These systems generally find application in protection of special hazards. The section also includes chapters on installation of fire extinguishers on the design of HVAC systems and on industrial process control.

Part IV (Chapters 20 to 23) includes fire protection problems usually assigned to the electrical engineer.

The encouragement and support of the Society of Fire Protection Engineers, which helped to initiate the project and sponsored the *Actual Specifying Engineer* series, is deeply appreciated. Without the contributions of the numerous fire protection engineers who wrote many of the individual chapters (while submitting to my editorializing), the book would not be.

The editors of *Actual Specifying Engineer,* from 1967 when the series began, through 1973 when it ended—Urban Reese (who had to teach me how to write), Bob Young, and Don DeMicheal (who kept me focused on the end when it seemed it would never come)—were a constant help.

Most important, one does not attempt an effort like this without an understanding wife whose generosity in providing me with "released time" from family duties is unending.

Part I

Codes, Standards, and Approvals

An Engineering Approach to Fire Protection

ROLF JENSEN, P.E.

"Fire protection? I don't have to worry about that; the insurance company takes care of it. Besides, this building can't burn."

Similar statements have been made many times—often by engineers and architects.

As a fire protection engineer, I find this alarming because no one is in a better position to foster greater protection of lives and property through improved fire protection engineering—at less cost—than the design professional. Fire protection engineering is a complex, many-faceted subject. While not limited to mechanical/electrical systems in buildings, the vast majority of fire protection engineering considerations are related to these systems. The engineer is, therefore, the member of the building construction team in the best position to comprehend the many and varied aspects of fire protection—and to make sure that the building is adequately protected. He may design the system himself or use the services of a specialist: the fire protection engineer.

Unfortunately, there are too few engineers involved in fire protection system engineering today to meet the building construction industry's needs.

The purpose of this book is to increase the knowledge of engineers in this vital area. The book will not make a fire protection engineer of the reader, nor is it expected to. A description of the formal education required for fire protection engineers is given later in this chapter. It will, however, equip the engineer to design better fire protection systems and, just as important, to recognize situations in which he needs the advice of the fire protection engineer. Either way, more buildings will have adequate fire protection. And, after all, this is the goal.

One of the major problems in fire protection engineering is that too seldom is much thought given, in the early building design stages, to the specific fire protection needs of the building and its occupants. As a result, many fire protection systems are designed merely to meet a local building code or a national fire protection standard. The flaw in this approach is that such codes and standards are general in nature; they may not protect against specific fire hazards.

Another major problem arises when fire protection systems are designed to meet the requirements of an insurance company or an insurance rating firm. Insurance companies primarily have the responsibility of rating what they see—of setting a price for insurance. They can tell the engineer what to do to obtain a lower insurance rate, but they have neither the time nor the motivation to tell him how to do this in the most efficient and economical way.

An example will illustrate the danger of providing fire protection that only goes as far as code or standard compliance.

Assume that you are the engineer working on a storage warehouse. Included in the design is a full complement of automatic sprinklers, installed in a system designed to follow sprinkler standard NFPA 13—the standard recommended by insurance people for such designs.

Unfortunately, standard NFPA 13 deals with sprinkler systems in general; it does not cover each kind of hazard condition specifically. If the engineer uses the average approach and simply follows the standard, the insurance company will have a comfortable basis on

McCormick Place, Chicago, lies in ruins after one of the most spectacular and most talked-about fires of recent times.

which to insure the warehouse at a reasonable cost. The question is: Will this sytem be overdesigned? Or, much worse, will the warehouse burn to the ground?

The engineer can go too far in either direction when following a code or standard. But overdesigned systems are expensive, and underdesigned systems can result in tragic losses of life and property.

Two examples from actual experience will illustrate what can happen with adequate and inadequate fire protection engineering.

The first concerns a large container manufacturing plant in New Jersey. In 1962 a small fire started in this plant, resulting in a complete loss of the building and its contents. What was thought to be a well-designed automatic sprinkler system, supplied with a more than adequate supply of water, served this building. The system conformed to standard NFPA 13. Materials in the building were stored twenty feet high. The fire started in an area used to store wax milk cartons. It spread so rapidly that it completely overtaxed the sprinkler system.

Tests made after the fire proved beyond a doubt that a properly designed fire protection system could have prevented much of this loss. The investigation was testimony to the fact that it is not proper merely to install in a building a sprinkler system with piping and water supplies loosely established on the basis of general piping schedules of the sprinkler standard.

The proper approach requires that a sprinkler system be specifically designed for hazards, with piping hydraulically sized, and with water supplies and flow rates calculated to produce a water discharge adequate for the size of fire which may reasonably be anticipated. Tests have shown that this approach will guarantee control or extinguishment of a fire such as this one.

The second example involves new buildings on an existing plant site. The initial design called for an "adequate protection system" for an addition, following existing codes. Part of this plan was to provide a new water supply connection, because the new buildings were to be located approximately one-quarter of a mile from the existing building. The plan originally required a connection to a city water main and construction of an elevated water tank at the new site.

Then the plant fire protection engineer made a simple suggestion—that a water main be extended from the existing building to the new buildings. This plan eliminated the need for the elevated tank, because a tank was already in place at the site of the old buildings. This change in plan provided an immediate benefit by keeping the fire protection system isolated from public water supplies which could deteriorate with industrial or domestic growth in the area. Two important savings were also realized: The new plan was $40,000 cheaper in initial investment, and elimination of the new tank brought about savings of $5,000 per year in maintenance costs.

Finally, this design allowed use of the fire mains to circulate water from the existing reservoir to air-conditioning equipment in the new buildings and saved the cost and operating expense of a cooling tower for air conditioning.

The only difference in the two approaches is that the first was based on existing national standards; the second added to those standards, to achieve the best fire protection for the least money.

These are two examples only. But they show that simple compliance with standard codes or regulations could not produce a design that accounted for all of the factors which should be considered for effective fire protection. It is a matter of concern that few engineers recommend better levels of protection or different methods of accomplishing protection, backed by a firm cost analysis.

In a sense, utopian design is still in the future. When an engineer comes up with a proper design for fire protection not meeting the specific level required by a standard or building code, there is often no way of convincing the building authority that he is on firm ground. But this is actually no more unusual in fire protection than it has been in structural engineering where similar rule of thumb factors of safety are imposed. It is one more good reason for close cooperation and teamwork between the design engineer and the fire protection engineer.

Many organizations regularly publish bulletins and information on fire hazards and fire protection. A description of this information and sources are given later in this chapter. Unfortunately, at the present time there is a notable lack of information in the form of basic textbooks for reference or teaching. Until these are developed it is hoped this will book will fill the gap.

What Is a Fire Protection Engineer?

A fire protection engineer is a highly specialized individual. The best way to define him is to describe the educational program to which he is subjected, and the business experience skills he develops after graduation. The curriculum of one of the two schools offering an FPE degree is shown in Table 1.1. This information is taken directly from the school bulletin. An examination of this curriculum shows that the undergraduate program in fire protection engineering is quite close to undergraduate engineering programs in other disciplines.

The Illinois Institute of Technology program requires two years of college-level physics, two years of college-level calculus and differential equations followed by a computer programming course, statistics, thermodynamics, heat and mass transfer, general and organic chemistry, statics, dynamics, strength of materials, fluid mechanics, AC and DC theory, and electronics.

All of these subjects form a base in the physical and engineering sciences. Specialty skills include risk management, building construction and structural fire protection, municipal fire protection and water supply, design of fire protection detection and extinguishing systems, basic phenomena of the combustion process, analysis and control of industrial process hazards and total cost related Fire Protection Systems Analysis. This part of the program is all technical, and is liberally supported with laboratory work.

A few liberal arts courses help on the economic side

Fireman on ladder is almost completely enveloped by billowing smoke.

Table 1.1.   Illinois Institute of Technology Fire Protection Engineering Curriculum

Freshman Year

Math (Anal. and Calc.)	Math (Anal. and Calc.)
Physics (Mechanics)	Physics (Heat and Wave Motion)
English	tion)
Inorganic Chem.	English
Graphics	Inorganic Chem.
Physical Ed.	Descriptive Geometry
Engrg. & Sci. Orientation	Physical Ed.

Sophomore Year

Math (Anal. and Calc.)	Math (Diff. Equations)
Physics (Elec. & Magnetism)	Physics (Modern Physics)
Prin. of Economics	Statistics & Dynamics
Phys. Chem (Thermodynamics)	Prin. of Economics
Fire Insurance Schedule Rating	Fire Insurance Schedule Rating

Junior Year

Fire Protection Engineering Lab.	Fluid Mechanics
Property Insurance	FPE Lab.
Structural Fire Protection	Casualty Insurance
Strength of Materials	Publ. Fire Prot.
Organic Chem.	Electronic Processes in Materials
Programming for Digital Computers	Heat & Mass Transfer
	American Constit. System

Senior Year

Liberal Studies Elective	Humanities Elective
FPE Lab.	FPE Application
Industrial Hazard Control	FPE Lab.
Electrical Hazard Control	Industrial Hazard Control
Fire Protection Extinguishing & Detection Systems	Fire Protection Extinguishing & Detection Systems
Electrical Circuit Analysis	Engineering Electronics

In practice, the fire protection engineer must be multi-disciplined, so that he can converse easily with engineers of all disciplines. Since the fire problem involves or affects structural design of buildings, he must have a basic understanding of building construction from both a design and a methods viewpoint. He must know interior finish materials as they affect the fire problem, and must know the fire hazard potential in the mechanical systems of the building. He must be conversant with the electrical engineer to eliminate hazards and assist in the design of proper overcurrent protection wiring systems and so forth. He must know water supply and hydraulic systems, since many of the building's fire protection systems use water. He must be well grounded in chemistry and chemical industry processes. He must know the insurance business and have a basic and broad understanding of economics.

Stated simply, the fire protection engineer must be a combination of civil, mechanical, electrical, and

of the picture. Economics, English, basic insurance analysis (since insurance is a prime economic factor in the design of fire protection measures) round out the course. The IIT program in Fire Protection Engineering originated in 1903 and has produced approximately 1,100 graduates.

The University of Maryland program has basic courses in mathematics, chemistry, physics, and the engineering sciences similar to the IIT program. The fire protection courses, although named differently, have the same scope as the IIT program. The Maryland program originated in the late 1950s.

Current annual output from the two schools is thirty per year. Graduates from both schools are given full professional recognition by the Society of Fire Protection Engineers.

The curricula of the two schools show that the fire protection engineer has a sufficient foundation in the physical and engineering sciences to be able to converse on an equal level with an engineer of any other discipline in working to the solution of a fire problem in a specific situation.

The building above will comply with almost any building code, and be favorably rated by most insurance people. But the metal grille, while esthetically pleasing, shows an absence of consideration for fire protection. The grille will prevent firemen from evacuating building occupants or from directing water into the building.

chemical engineer with a deep understanding of the fire problem. Because this is such a demanding background for any discipline, many practicing engineers have resorted to handbook engineering under such stress.

Fire protection engineers have been used primarily to analyze existing hazards in buildings and processes. Only rarely have they been used to attempt economic engineering solutions to the elimination or control of these hazards. This is a backward approach. Imagine the hodgepodge of buildings we would have today if every architect or civil, mechanical, structural or electrical engineer worked for a city building department rather than in his proper place in an architectural, contracting, engineering, or construction firm?

Fortunately, in the last ten years an increasing number of fire protection engineers are being employed by private industry and government, and several have found their way into the consulting business. Here, the creative skills of the fire protection engineer may be used to great advantage when he is working at the level where design originates. Clearly his education qualifies him for both efforts. A good number of fire protection engineers practicing in various segments of their field today are registered professional engineers. They have gone through their required examinations in the same manner as other engineers. There is also a professional society—a member of the Engineers Joint Council known as the Society of Fire Protection Engineers (SFPE). This book would not have been started were it not for the sponsoring effort of the SFPE, whose members are contributing authors.

The SFPE functions in much the same manner as the founder societies, has similar grades and requirements for membership, and sponsors a bimonthly publication, *Technology Reports,* which contains up-to-date technical articles. This magazine is currently the highest technical level magazine in the fire protection engineering field and regularly contains articles of significant interest, covering new design methods and techniques or applied or basic research.

Information Sources

Undoubtedly, the most productive source of information in the field of fire protection is the National Fire Protection Association.

This nonprofit technical organization is an association of individuals interested in the fire problem. The NFPA maintains, publishes, and distributes a complete series of standards (National Fire Codes) covering almost all areas of fire protection. These standards are developed by technical committees through a process

comparable to that utilized by ASTM or USASI (see Appendix). NFPA also publishes some information on fire statistics, operating and training manuals, inspection manuals, and fire research. It publishes the NFPA Handbook of Fire Protection, an excellent general reference text.

The American Insurance Association publishes technical bulletins describing methods of detecting or controlling specific processes, hazards, methods of approach for building structural protection, and similar subjects. This organization publishes the National Building Code and occasionally publishes research reports which eventually lead to recommended good practices or standards in areas not covered by the NFPA.

Underwriters Laboratories, Inc., publishes seven annual lists: Building Materials; Fire Protection Equipment; Electrical Appliance and Utilization Equipment; Construction Materials; Hazardous Location Equipment; Accident, Automotive, and Burglary Protection Equipment; and Gas and Oil Equipment. These are listings of manufactured products which have been tested by UL and found to meet their requirements. The products are inspected under the UL factory inspection and label service program. UL also published bulletins of research, data cards, and standards. The first two cover a wide variety of subject areas within the broad range of fire protection and safety. The latter describes UL test requirements.

Industrial Hazard Information

The Factory Insurance Association publishes a series of recommended good practices for protection from many industrial hazards. The FIA also publishes reports of sponsored research conducted by them, not all of which are readily available.

The Factory Mutual Engineering Division publishes a list of Factory Mutual Approved Fire Protection Equipment similar in intent and scope to the combined UL lists of Fire Protection Equipment, Hazardous Location Equipment, and Gas and Oil Equipment. Factory Mutual, through subsidiary Factory Mutual Research Corporation, publishes a number of research papers on many areas of sponsored research. The availability of these is often limited to the sponsoring organization. This organization also publishes standards covering its equipment approval operations and other standards covering fire protection design recommendations for industrial processes and so forth.

The Factory Mutual Engineering Division staff has authored the *Handbook of Industrial Loss Prevention,* a compilation of Factory Mutual recommendations and

Fires such as this can be prevented by tailoring the fire protection systems to specific building needs.

methods for protection of industrial processes, and publishes an Engineering Data Book, available on a subscription basis which includes updating.

AGA Test Program

The American Gas Association publishes a list of Approved Materials for equipment which has been tested and found satisfactory by the AGA Laboratories. It and the American Petroleum Institute also publish a number of bulletins covering fire protection recommendations for certain specific problems in the petroleum industry.

Technical data sheets are available from the American Chemical Society, the Manufacturing Chemists Association, the Compressed Gas Association, the American Petroleum Association, and from various manufacturers.

Research reports are published by a number of organizations, in addition to those previously mentioned, including the US Naval Research Laboratory, the Department of Defense, Illinois Institute of Technology Research Institute, Southwest Research Institute, and others.

Building Codes

WILLIAM PRITSKY, P.E.

"In the case of collapse of a defective building, the architect is to be put to death if the owner is killed by the accident; and the architect's son, if the son of the owner loses his life."

Building codes are tough today, but not as tough as that severe punishment administered under Hammurabi, king of Babylon around 2100 B.C. His is the earliest building law on record.

Though things have brightened considerably in the last 4,000 years or so, the need to comply with building codes, and to assure compliance by others (contractors and suppliers, for example), is just as important.

Almost every building and building system must comply with a building code—and approximately 75 percent of code requirements relate to fire protection.

Four Model Codes

There are four nationally recognized model building codes. The first code written, now known as the National Building Code, was recommended by the National Board of Fire Underwriters in 1905. It was published by the American Insurance Assn.

In 1927 the Uniform Building Code of the International Conference of Building Officials was published. The Southern Building Code was developed by the Southern Building Code Congress in 1945. The basic building code of the Building Officials Conference of America was published in 1950.

There are many state, county, and city codes, as well as standards and regulations of federal agencies that deal with buildings and systems. Most are patterned after one of the four national model codes and are important to the design engineer.

Most areas also have fire prevention and zoning or-dinances in addition to the codes mentioned. The fire prevention ordinance sometimes contains requirements for installation and maintenance of protective systems, such as fire alarms, standpipes and sprinklers. Zoning ordinances may establish fire limits and thereby influence the basic limits on heights, areas, and type of construction.

In addition, in many areas code manuals describe acceptable equivalent methods of complying with the code. Obviously, such documents are of immense value to the engineer. They are all available through a city building department or similar agency.

Two of the four model codes have stated purposes.

The purpose of the National Building Code is "to provide for safety, health, and public welfare through structural strength and stability, means of egress, adequate light and ventilation and protection to life and property from fire and hazards incident to the design, construction, alteration, removal or demolition of buildings and structures."

The Uniform Building Code contains a similar statement. The other two model codes do not but are written to meet essentially the same purpose. When revisions or amendments to model codes are contemplated by the sponsoring organization, the basic test applied to such changes is whether or not they meet the code purpose.

Revisions

Revisions to model codes are made in a prescribed and formal manner, following procedures set forth by the code-writing organization. Requests are submitted in writing, with substantiating background material. In three of the model code organizations, requests are

studied by a code revision committee and voted on at a
regular annual meeting of the code organization. Code
revisions are subject to both technical review and
political pressures.

Proposed revisions to the National Building Code are
studied by the staff of the American Insurance Assn.
or its building construction committees. These groups
evaluate proposals in light of national standards and
the lessons learned from actual fire losses.

They use studies of fires and fire research by organizations such as the National Fire Protection Assn., the
American Insurance Assn., National Bureau of Standards,
Factory Mutual Research Corp. and Underwriters
Laboratories, Inc. These studies yield data on the fire
hazards of buildings, occupancies and materials, and
the means by which fires spread in buildings. As knowledge in these areas increases, new code requirements
are developed.

Codes must be dynamic instruments or they soon
become obsolete. They must keep pace with changes
in occupancy hazards, with changes in services and
with new materials that can affect combustible loading
or potential fire severity and spread.

New materials and methods of construction must
be recognized and regulated appropriately. To provide
protection of life and property, codes must keep
abreast of advances in fire protection engineering. Unfortunately, most codes fall short of the ideal—either
in the way they are written or applied.

Most codes only broadly treat the wide variation of
fire hazards in a building. Occupancy classifications
mostly related to "people loading" or preservation
of the structural integrity are only generally related to
fire loading. Protective systems commensurate with
the hazard are required only in isolated instances.

An Alternative Approach

Seldom does a design engineer objectively attempt
an equivalent method of meeting a code requirement.
Code officials seldom encourage such an approach,
since it is unreasonable to expect code enforcers to
initiate alternative approaches. This effort must start
with the engineer, the architect and the equipment
and material suppliers. To be valid, proposals must be
objective.

While there is no guaranteed successful method of
obtaining acceptance of an equivalent design, the following is probably best. Begin with the responsible code-
enforcing official—usually the head of a city building
department. Discuss the proposition with him to find
out if it is reasonable and feasible. He will then need a
formal submittal, related to a specific building design,

which is usually made to some kind of appeals board. The submittal should present the case objectively and include specific proposals. Applicable fire records, test data, established standards and other published literature should be used as supporting references.

Sometimes a formal code revision is needed and justified. Under such conditions a stronger case must be established. One should normally anticipate the need for legal adoption by a governing body, which may also entail public hearings. Cases involving equivalent design may take two to four months to resolve, but if legal adoption is necessary, it may take longer.

Area Limits

One of the more important code requirements concerns area restrictions. The basic reason for limiting areas is that large areas make large fires. With sufficient combustibles, a fire generally will spread until contained by physical barriers or until it is extinguished. Some buildings are so large or so tall that fire department hose streams cannot reach a fire from outside.

Basic area limits are influenced by type of construction and nature of occupancy. The classification of buildings by construction type is related both to combustibility and ability of structural members to withstand fire exposure for specified periods of time. Classifications vary from wood frame (combustible with no significant fire resistance rating) to fire resistive construction (noncombustible with fire resistive ratings of up to four hours).

Since the concern is to limit fire spread and building collapse until the fire can be extinguished, larger areas are generally permitted for buildings having greater fire endurance and accessibility from more than one side. Larger areas are permitted in some codes for low-hazard occupancies in a given construction type.

Significant area increases for the installation of automatic sprinklers are often overlooked. Area limit tables dictate a certain type of construction for the size of building contemplated, but less expensive construction may be used with automatic sprinklers. Location relative to streets will also significantly affect the area allowed.

Remember that area increases are additive. Add the

Fire Protection—Whose Responsibility?

Courts have ruled consistently that approval of a building design or building specification does not relieve the engineer or architect of his responsibility to comply with the law and the provisions of a building code.

If a building design feature is found to be a proximate cause of a fire, most courts rule negligence per se and instruct the jury to return a verdict of guilty.

The following case involves an architect and contractor but is illustrative of an important point that can affect mechanical/electrical engineers in their designs and specifications.

During the sixties, a large hospital was designed for a Midwestern city. Drawings were submitted to the local code authority for review and were approved. Final inspection was made after construction, in order to issue a certificate of occupancy.

Two major deficiencies were discovered. One involved construction of wall sections surrounding vertical pipe shafts. The walls were of an area that required them to be of two-hour fire resistive construction. Although the walls met this requirement on the corridor side, they did not on the shaft side.

The second deficiency involved wall surfacing materials used throughout the corridors in the patient room areas. The material had a flame spread rating of more than 200, based on unofficial tests, but the code required a maximum flame spread rating of 25 for these areas.

The extent of the wall surfacing areas and the number of shafts involved were great enough to require an additional $1,500,000 to correct the deficiencies. Corrections had to be made before the code authority would issue an occupancy permit.

The hospital subsequently filed suit against the architect, contractor, and wall-surfacing material supplier. The basis of the suit was that the original contract with the architect required that the building be designed and constructed in accordance with all applicable provisions of the code. The architect was paid for supervision of construction, an important point in the suit.

The architect had issued general orders and some special orders to the general contractor and the material suppliers to construct the building and supply materials in compliance with local codes. This also was an important point.

The case was tried in early 1967, and the jury found in favor of the hospital. All three defendants were held liable. The extent of individual liability was determined in a separate trial.

percentage increases and apply them to the tabular area.
This gives allowable area increase in sq. ft. For total
allowable area, add the tabular area.

Fire Limits and Fire Spread

Reducing the possibility of fire spreading from one
building to another is a function of codes. Each building
owner has the right to expect that his building will not
be jeopardized by fire in a nearby building. Codes treat
this problem by requiring exterior walls to withstand
fires for certain time periods and by requiring protection of wall openings. The amount of space between
buildings is also a factor.

Early building codes established areas in cities known
as fire limits or fire zones. Within these zones all building walls were required to be masonry of minimum
thickness to reduce fire spread.

Masonry exterior walls originally were required in
such areas, but newer materials may be used. Exterior
walls may be of any construction of noncombustible
materials having a two-hour fire resistance rating. Codes
generally allow larger areas outside fire limits, because
the threat of a general conflagration is small, and large
areas are needed for commercial and manufacturing
activities.

Accordingly, codes contain provisions for unlimited
areas outside the fire limits. If the fire limits are properly delineated, the public is protected and practical
business needs are served. Unfortunately, many cities
do not properly lay out their fire limits. They may desig-
nate all business or commercial areas as fire limits,
which, in effect, negates the code provisions for unlimited area.

Protection of Interior Wall Openings

Chapter 6 describes differences between fire resistance
ratings for structural components and for fire doors
and windows. An understanding of this difference—
especially with respect to temperature transmission
during a fire—is basic to the design of proper protection
of openings, including those for building m/e systems.
Structural components are rated on both structural
integrity and temperature transmission during fire exposure; opening protectives are rated only on structural
integrity. Thus, it should be obvious that any opening
in a fire barrier is a weak link, whether protected or not.

Codes generally specify an hourly rating for opening
protectives related to the fire resistance of the fire wall
on which it is placed. However, rarely does a code
recognize that there are equivalent methods of protecting openings, such as water curtains or window sprink-

lers. As a result, these methods are not often employed.
The equivalency of such methods depends on reliable
design and proper maintenance.

It is also significant that the use of an opening as a
means of egress may be lessened where Class A opening
protectives are used. Codes do not provide for the use
of less fire resistive opening protectives where the fire
hazard of the occupancy is low.

Fire doors are obviously useless unless closed during
a fire—yet they often interfere with normal traffic
patterns in a building. When they do, wedges or door
blocks soon appear. The solution is a simple one—provide
hold-open devices and automatic door closers by specification at the design phase.

One of the greatest conflicts between code requirements and building design needs is the desire of engineers
and architects to traverse the building vertically with
openings. This need relates to aesthetics, natural light,
and to the building's mechanical and electrical systems.

To the codes and to the fire protection engineer,
vertical openings are potential chimneys—and the fire
record confirms this. Many fire-safe buildings are constructed each year with protected vertical openings.
Fire dampers, sprinklers at the top of shafts and fire
rated shaft doors do much to eliminate the problem.

A code should be thought of as a permissive document, not a restrictive one. Its purpose is to protect
the public—not prohibit design concepts. Of necessity,
a code is general in nature. It cannot be written for
a specific building. A general rule often followed in
code interpretation is that anything not specifically
prohibited is permitted.

References

Basic Building Code, Building Officials Conference
of America (1965).

National Building Code, American Insurance Association (1967).

Southern Standard Building Code, Southern Building Code Congress (1965).

Uniform Building Code, International Conference
of Building Officials (1967).

Building Codes, Their Scope and Aims, American
Insurance Association (1957).

Life Safety Code, National Fire Protection Association (1967).

Design and Construction of Building Exits, National
Bureau of Standards (1935).

Guide on Building Areas and Heights, National Fire
Protection Association (1965).

Recommended Method for Laying Out Fire Limits,
American Insurance Association.

Design Approval
Authorities and Agencies

BOYD A. HARTLEY, P.E.

Admittedly, the subject of fire protection systems approval is a complicated one. Misunderstanding usually arises for two reasons. The design team may be 1) unaware that there is more than one authority having jurisdiction over design, or 2) be unaware of the authority's sphere of influence or limits of jurisdiction.

In the design approval of any building system or component, three essential factors are involved. The first is the design itself. The second is some organization, office, or individual having approval responsibility. The third is a law, ordinance, code, standard, or recognized good practice that the authority uses as a basis for design review. This chapter will discuss the second and third factors.

Authority Having Jurisdiction

The organization, office, or individual having approval responsibility is referred to in fire protection circles as the "authority having jurisdiction." The term is widely used in all standards of the National Fire Protection Assn. Unfortunately, it is not always understood. One reason is that there may be several authorities having jurisdiction, each for a different purpose.

The most important authority having jurisdiction is the governmental authority. Since it is the only authority having law enforcement powers, it must always be consulted. A governmental body with legislative powers appoints the governmental "authority" for enforcement of codes, laws, and standards. If there were a single legislative body and a single "au-

thority having jurisdiction," the business of building would be simplified. But such is rarely the case.

Codes and laws governing building construction, including fire protection, may be adopted at several legislative levels. The municipal government (city council), county government (board of commissioners), state government (state legislature), and federal government (Congress) are all involved in building legislation, including the appointment of authorities with jurisdiction at each level. The result is that the building design team must usually satisfy several government enforcement bodies, whose areas of authority often overlap.

Occasionally, a governmental body will adopt a code and delegate its enforcement authority to a nongovernmental organization. There are still large areas in some of our most heavily populated states where the power to inspect and approve installation of electrical equipment has been delegated to organizations supported by or under the control of the insurance business. For example, in certain areas of New York State, electrical inspection and enforcement is by the New York Board of Fire Underwriters. This practice, common in the past, appears to be slowly evolving into enforcement by municipal officials.

Some communities have no fire protection laws, lack laws in certain areas of building design, or have inadequate laws. In such communities, local officials such as the fire chief or building inspector make recommendations to property owners for improvement of property or elimination of hazards based on recognized good standards or what the official believes the law should be. In almost all such cases, the local official is doing his best to reduce the fire hazard of his community without having the legal backing of laws adopted by the appropriate governmental body. The amount of actual law enforcement authority in these cases cannot be spelled out because of the vagueness or complete lack of enabling legislation in the municipality.

Another situation is one where laws and codes duplicate each other or overlap in jurisdiction. The laws of a municipality may require that a building design be submitted to both the building inspection department and the fire prevention bureau, the latter usually being a division of the fire department.

Because of poorly worded laws, there may be overlapping of requirements between the two agencies. However, in general, the building inspection department will be responsible for enforcement of the building code (the laws governing building construction), while the fire prevention bureau will be responsible for enforcement of the fire prevention code (the laws governing building occupancy). Each enforcement authority considers problems peculiar to its own jurisdiction but normally will also consult with the other enforcement authority.

Code Problems in Design

Designers who have a basic concept of what the hazards of fire are, how a fire may affect the building's equipment and materials, and how a fire may affect the occupants rarely find themselves in major trouble with fire code requirements. Such designers also concern themselves with the ways and means these occupants might escape. But designers who have the idea that the fire regulations are a hindrance to their creative talent are likely to have trouble. They use their energies and imagination in devising layouts which may comply with the actual wording but not necessarily the intention of the code. For example, the need for two separate and distinct stair systems should be obvious. However, the necessity of one being *remote* from the other is stated in the code as "remote as practical," and this factor is often ignored in design.

This stairway problem is well illustrated in a May 1967 article in *Fire News* (published by the National Fire Protection Assn.) which discusses the tragic Cornell University dormitory fire of April 5, 1967. The article states:

> Inadequate and substandard escape facilities were principally responsible for the death of eight students and a faculty member in the fire April 5 at a Cornell University dormitory, Cayuga Heights, New York.
>
> While the building had two avenues of escape to the outside, they were located in close proximity to each other, contrary to the requirements of NFPA No. 101, the Life Safety Code, which specifies that two or more building exits be widely separated.

A procedure which may save a designer both time and money is to submit plans for permit before contracts are let. Such plans may be processed through all plan examiners and necessary corrections noted so that corrections may be made and accepted prior to bidding. Such a procedure eliminates many costly extras involved in changes after contracts are signed. It also increases the chance of achieving a completed structure free of violations.

It is important to note that approval by a code authority does not relieve a designer of legal responsibility to comply with the codes. This has been supported by several legal decisions; serious liability may be incurred if a code design violation is judged to be the proximate cause of a fire.

Fire prevention bureaus of most fire departments are vitally interested in the provisions of a building code. They are especially concerned with exits, interior finish, protection of vertical openings, and other important features of construction relating to the spread of fire and the safety of the occupants. Designers should submit complete process descriptions with plans or have a company representative who is completely familiar with process operations discuss the problem with the fire prevention bureau. This helps the bureau representative arrive at a proper solution. For example, information vital to a decision by the enforcement authority includes the type and amount of flammable liquid, where and how it is to be stored, and handled, and what protection features are to be provided.

Be cautious of installation problems. For example, a workman who scratches the finely machined flange of an explosion-proof electrical fixture may nullify the intended function of the fixture. This could seriously affect the designer when construction supervision is a part of his contract.

Insurance Interests

An important authority having jurisdiction is generally recognized to be insurance interests. Unless they have been officially designated by a governmental body as having law enforcement authority, the authority of insurance interests or insurance supported organizations is only advisory in nature.

Since their recommendations are advisory, they may be ignored by designers. However, their advice should not be rejected without consideration. At the least, their advice is intended to reduce the fire hazard problem of a building and should be beneficial to the property owner. It should also be understood that disregarding the advice may result in a higher insurance rate on the building or, in drastic cases, inability to obtain insurance.

Organizations having an interest from the insurance point of view fall into three categories. They are insurance companies; rating organizations; and advisory organizations.

Insurance Companies

Insurance companies rate insurance on a property or building, issuing an insurance policy to the property owner. In most cases, an engineer's contact will be with individual insurance companies; however, individual companies often band together, pooling their assets, to write large amounts of insurance on large industrial properties or in specialized areas of business activity usually through associations organized for that purpose.

The most commonly encountered organizations or groups of insurance companies are the Factory Insurance Association and Factory Mutual Insurance Association. The latter organization is commonly referred to as Factory Mutuals. An example of an insurance company group formed to write insurance in a specialized area is the Oil Insurance Association.

Rating Bureaus

The most commonly encountered insurance interest is the group of organizations known as rating bureaus. Because the insurance business is subject to state rather than federal regulation, there exists in each state an organization, supported by the insurance companies doing business in that state, with the responsibility for filing standardized fire insurance rate-making procedures with the regulatory officials of the state.

The individual rating bureaus in most states have recently merged into a national insurance-supported organization named the Insurance Services Office and are now state offices of the parent organization. They also have responsibility to determine insurance rates on individual properties within that state for the members and subscribing insurance companies.

A major exception is the state of Texas where insurance rates are established by a state-supported bureau; the insurance-supported organization in that state falls in the general category of advisory organization.

Rating bureaus have many other responsibilities. Among them are inspection of insured properties and buildings to determine the degree of fire hazard and inspection of municipalities to determine the adequacy and reliability of municipal fire protection facilities. The Factory Mutuals and some other insurance companies normally do not subscribe to these rating bureaus, but operate separate rating organizations.

From the viewpoint of a designer, the insurance rating organization having jurisdiction should always be consulted. The organization may not have jurisdiction over insurance rates on a specific property, because the property either is self-insured or non-insured or is insured by some organization not a member of the bureau. The bureau engineer will so advise you and will usually state that he cannot comment on the plans. In those instances, it will be necessary to consult for advice with the particular insurance company or organization writing the insurance.

Designers should consult with the insurance interests at the earliest practicable date. You should obtain from the building owner the name of any insurance company or companies that will be covering the property. During all plan stages in the design procedure, a designer should obtain answers to specific questions and have plans reviewed by insurance authorities.

A hypothetical situation in which insurance interests were not consulted until final definitive drawings had been prepared illustrates what can happen if this is not done. After working drawings were prepared, a designer visited the insurance interests with a set of architectural drawings and preliminary sprinkler layouts. The insurance representative, having no prior knowledge of the project or sight of the building plans, recommended "minor" changes in the details of the plan. However, a certain product, piled twenty-two feet high, was to be stored in a warehouse section, and proper fire protection required that the sprinkler system be a hydraulically balanced system with two fire pumps and a million-gallon ground storage reservoir.

A rough estimate of the cost of the recommendations indicated that compliance would add 20 percent to the total project cost. The alternative was prohibitive insurance rates. There were other alternatives to the recommendations. Reducing the height of piles to ten feet was one solution, but possibly one-half the height of the building would remain unused, which would be uneconomical.

Another possibility was modifying the design of the structure to increase floor area and reduce the height of piles in the warehouse without reducing the total storage volume. This would mean drawing new architectural plans and having them approved. A third alternative would be to spend the additional money and comply with the recommendations.

Many buildings in existence today would have a vastly different shape and form without any additional expense to the owner if the insurance authorities had been consulted in the initial planning stages. A savings on insurance premiums would have resulted and would have been a significant continuing cost factor if the design of particular buildings had been modified in the early design stages while still providing an efficient building with adequate fire protection facilities.

Advisory Organizations

The third category of insurance interests are the advisory organizations. As they are normally advisory to either the rating bureaus or the insurance companies who are members of the specific advisory organizations, designers will rarely have contact with them.

An example of an advisory organization operating on a national scale is the American Insurance Association. It was formed January 1, 1965, by a merger of the National Board of Fire Underwriters (operating in an advisory capacity in the fire insurance field), the Association of Casualty and Surety Companies (operating in an advisory capacity in those fields), and the American Insurance Association, an earlier organization.

The American Insurance Association is an organization supported by capital stock insurance companies through either membership or subscription. Its primary responsibility is as a service organization for the member insurance companies. It is also an advisory organization to the rating bureaus. Its activities include many that can be considered a public service.

There are also advisory organizations serving the mutual insurance companies, of which the best known would be the American Mutual Insurance Alliance, headquartered in Chicago.

Organizations having an insurance interest in the property are in a position to assist both industry and the designers. By commenting on fire protection measures, they can help reduce potential loss of life and property due to fire and help reduce insurance rates. However, their interest is limited to the insurance aspects of the property. The economics involved in choosing one recommendation over others available is the responsibility of the property owner or his designers. Insurance interests should not be expected to take the place of the engineer, whose detailed analysis may often be needed to achieve the desired protection at the minimum cost to the property owner.

Basis for Decisions

The third factor in the approval process, mentioned at the beginning of this chapter, is the basis upon which an authority having jurisdiction makes his decisions.

At the governmental or law enforcement level, the basis for decision will usually be a law or code adopted as a law. A few governmental bodies still attempt to write their own building codes and other laws. But the general trend is to adopt codes and standards developed by organizations having broad background and experience in the particular field involved.

Some examples of standards available for adoption and the organizations promulgating the codes or standards are:

1. Building Code Requirements for Signs and Outdoor Display Structures, A60, United States of America Standards Institute (formerly American Standards Association).
2. Uniform Building Code, International Conference of Building Officials.
3. BOCA Basic Building Code, Building Officials and Code Administrators International.
4. National Building Code, American Insurance Association.
5. Elevator Safety Code, sponsored by the American Institute of Architects, the American Society of Mechanical Engineers, and the National Bureau of Standards, through the United States of America Standards Institute.
6. The National Electrical Code, National Fire Protection Association.

Life Safety Code

The Life Safety Code (NFPA 101) is probably the most widely enforced code in the field of fire protection. It is revised through the Committee on Safety to Life of the NFPA with membership through appointment by the NFPA Standards Council. Revisions to the code may be proposed by any individual or organization having an interest in the subject. Proposed revisions should be sent to the NFPA. They will be forwarded to the committee and processed for adoption through the membership of the NFPA at a regular meeting.

However, adoption by the NFPA does not give a new edition of the Life Safety Code any force at law. It is a general rule of law that legislative bodies cannot delegate their legislative authority. Therefore, each successive edition of the code must be considered for adoption in each jurisdiction. Only when adopted by the governmental authority concerned does it become the basis for enforcement decisions by the authority having jurisdiction.

This statement is also true of many other codes and standards available for adoption by governmental bodies. Until such adoption, they are merely recommended standards or standards of good practice and are not laws. If a governing body does not keep abreast of the times, it is possible for the law in that community to be based on a code or standard many years out of date rather than on the current edition.

Therefore, it is essential that every engineer become familiar with the law in a particular municipality. Reliance on a current edition of a code for design purposes may prove a waste of time and effort if the law where the building is to be constructed is based on a previous edition of a code. The opposite is also true. Unless an engineer is constantly working in a particular area, he is not likely to keep abreast of the latest revisions to applicable standards. Too frequently the engineer will go to the office files or library, obtain the most recent copy of a particular standard in the files, and base his design on that edition of the standard.

Again, time and money is wasted if he finds, when he submits the plans for review to an enforcement authority, that the most recent edition in the office files is some five or ten years out of date.

Operating Rules and Practices

It should be recognized that in implementing any code or standard, the enforcing or review agency may adopt certain operating rules or practices in order to better define specific application of the code or to establish uniformity of interpretation. Such rules and practices are often available in writing. In the fire protection field, the standards of the National Fire Protection Association are often the only ones available and are usually the most widely used. All members and associate members of the National Fire Protection Association are advised of changes in any of the standards and are given the opportunity to present their views at a regular meeting of the association.

In addition, the literature received regularly by the membership will frequently contain information on codes and standards published by other organizations. Every engineering firm, as well as individual engineers, should become associate members of the National Fire Protection Association. The cost is nominal and well worth it. ($30.00 per year in 1975.)

NFPA Standards

Standards of the National Fire Protection Association, as well as other nationally recognized codes and standards, are normally the basis on which insurance interests will establish their recommendations.

When enforcing a code, it is often necessary for the authority to recognize or give approval to the use of a particular device or piece of equipment. The responsibility of the authority having jurisdiction is to approve generally or to accept only a material, process, construction, or installation with which he is either personally familiar or knows has been investigated by an agency he believes is both unbiased and reliable.

Such investigative or testing agency may be (and on rare occasions has been) established by law enforcement authorities at either the municipal or state level. In most cases, the authority having jurisdiction, whether responsible for law enforcement or only for recommendations, will rely on tests conducted by nationally recognized laboratories having a reputation for unbiased and impartial treatment of all products and manufacturers.

In the fire protection field, the most important nationally recognized testing laboratories are Underwriters Laboratories, Inc., the Factory Mutual Laboratories, and the American Gas Association Laboratories.

In connection with testing agencies, it should be kept in mind that approval of materials or construction is vested in the authority having jurisdiction. Labels and recommendations of recognized testing agencies are factors to be taken into account by the authority having jurisdiction but may, if desired, be rejected and approval based on other factors considered pertinent by the enforcement authority.

UL Product Approval Methods

JACK A. BONO, P.E.

Underwriters Laboratories, Inc., offers several services to assist the designer in the selection of fire-safe products.

The best-known example is the UL label. Products that meet UL's fire safety specifications are identified with a distinguishing label or marking. The label may include the fire rating or flame-spread classification the product earned in actual test. This label serves as an assurance that the properly specified materials reaches the job site and can be identified.

In its various lists, UL provides the names of manufacturers who make products that have been tested. Lists of fire-safe products are the results of thousands of hours of concentrated engineering effort by leading fire protection engineers, who test and evaluate various products and constructions for fire performance. Since the lists reflect only successful tests, there is no way, of course, to show the number of unsuccessful tests. Yet the unsuccessful tests have often led to the development of successful test designs.

To understand fully the value of the UL labels and listings, one should be aware of the objectives and purposes of the organization, its history, and the work done by its technically trained staff.

Underwriters Laboratories, Inc., was founded in 1894 to assist the insurance industry by establishing standards and then certifying products, materials and systems that complied with those standards. Insurance companies made use of the laboratories' findings in evaluating property insurance risks.

For the first twenty-two years of UL's existence, the capital stock property insurance companies contributed to the financial support of the laboratories.

After the insurance interests ceased financial support, the UL findings were made available, on a completely independent basis, to all interests concerned with public safety from fire, casualty, and theft. The organization is operated for service, not profit.

Although UL's early work was concerned primarily with electrical products, the scope of its operations expanded through the years. Today, as stated in its charter, its *raison d'être* is to ascertain, define, and publish standards, classifications, and specifications for materials, devices, constructions, and methods affecting such hazards, and other information tending to reduce and prevent loss of life and property from fire, crime, and casualty.

UL started as a one-room operation. It now has five major testing stations, in Chicago; Northbrook, Illinois; Melville, New York; Santa Clara, California; and Tampa, Florida. These stations are staffed by more than 350 graduate engineers and house the comprehensive testing equipment necessary for routine tests and special investigations. The organization also maintains an inspection force of more than five hundred inspectors and two hundred inspection centers throughout the world. In 1973, these inspectors conducted more than 254,000 inspections at 6,000 manufacturers' plants.

How Investigations Are Made

UL follows a number of fundamental procedures in its investigations for fire safety. The investigations are not undertaken simply on the basis of a manufacturer's

sending in his product. After a submittal is made, a
UL engineer goes to the factory where the product is
made to observe the manufacturing process and inspec-
tion-control procedures and to collect data. The material
to be tested is then packaged, sealed and initialed by
the UL representative for shipment to one of the testing
facilities.

This procedure insures that the test material is truly
representative of intended production. The data recorded
at the manufacturer's plant is later used in determining
the specifications under which the product will be
produced and inspected in order to be eligible for UL's
follow-up services.

Prototype products may be investigated. When this
is done, UL representatives will first make production
inspections to see that production products comply
with requirements. If the material is tested as part of
an assembly, close attention is paid the construction
of the assembly for test purposes, and an agreement
on the nature of the assembly is reached before the
test is conducted.

No test of an assembly is made just because it is
proposed by the manufacturer. UL does not exist solely
to test what a manufacturer proposes. It also recognizes
that there are numerous ways in which a specimen can
be assembled to improve its test performance; so before
an assembly is constructed, UL determines if a proposed
improvement for the test can be achieved in actual
construction.

After the successful conclusion of an investigation
of new or unusual products, a report of the results is
sent to one or more of the four UL engineering councils.
Councils currently in operation are those concerned
with the prevention of fire, burglary, casualty, and
electrical hazards. Each council is made up of persons
exercising authority in the interest of public safety, plus
insurance representatives who have special skills in the
product-installation area under jurisdiction of a particular
council.

The councils review the conclusions of the investigations before the reports are released. Negative comments from any council member are circulated among
other experts in the field for review and opinion. The
members of the councils are also invited to guide UL's
staff in other phases of its activities.

Standards

UL standards provide specifications and requirements
for the construction and performance (under test and
in actual use) of systems and products submitted to the
laboratories. The organization makes it possible for
manufacturers to participate in the development of
standards through its industry advisory conferences
of manufacturers' representatives. During 1973, nearly
one hundred such conferences were held to discuss
proposed new standards, as well as to bring existing
standards up to date.

In addition, drafts of standards are circulated among
the appropriate engineering council, members of committees of other organizations (such as the National
Fire Protection Association and American Society
for Testing and Materials), interested consumers, and
governmental representatives. Comments are solicited
from each group.

UL standards of safety are not intended as quality
standards for an industry, but it is difficult to draw a
sharp distinction between the two. In many cases,
UL has been asked to enter a product area in which
safety standards left much to be desired. In such cases,
the organization has been able to effect an improvement
in the overall safety and sometimes the quality of the
product. More than 100,000 copies of UL standards
were distributed to engineers, clients, inspection authorities, and others during 1973. These standards
covered three hundred different products.

Follow-up Services

UL's investigation procedure does not end with a report's acceptance by the engineering council and the
subsequent issuance of the report. Following these
steps, an inspection manual describing the tested product
in detail is prepared. It is given to the UL inspector in
whose area the manufacturer's plant is located. The
inspector periodically observes the manufacturing
process to make sure that the product continues to be
manufactured within the specifications developed from
the original test specimens.

The UL label or another agreed-upon marking on
the product or on its container means the product has
been produced under the factory follow-up program.
The UL labels are issued by the local inspector as needed
by the manufacturer.

The UL inspector frequently checks the efficiency
of the manufacturer's inspection program in the factory.
If the inspector, by conducting examinations or tests,
finds that the product is not being manufactured in
compliance with the requirements developed, the manufacturer must either correct the situation or else remove
the UL label from the product.

In many cases, follow-up examinations and tests are
conducted at UL facilities as well as at the factory. These
tests are made on samples of labeled products selected

in the plant or purchased on the open market and serve
as an effective countercheck on the factory inspection
program.

Acceptance and Benefits of UL Findings

UL-listed products enjoy an almost universal acceptance
by inspection authorities having jurisdiction over the
products. Though acceptance of UL-labeled products
by all inspection authorities is certainly not guaranteed,
a vast majority of authorities do recognize them. The
advantages and benefits of specifying UL-listed products
include the following:

- Assurance that the product specified will meet UL
  requirements for listing, because of the ongoing
  inspection procedure.

- Wide acceptance of labeled products by inspection
  authorities and users.
- Quick determination, by using UL lists, of manufacturers who produce labeled products.
- Assurance that the product has been extensively
  tested and then listed only after evaluation by many
  experts.
- Confidence that Underwriters' Laboratories, Inc., is
  absolutely impartial at all times.

The exacting examinations and tests, the attention
to detail, the service and the sense of obligation required
by UL's unique position in the field of public safety
have earned the respect and confidence of both consumers and authorities who are responsible for public
safety. In the final analysis, however, it is through inspection and identification of products as they are
manufactured that Underwriters Laboratories imple-
ments its aim of "testing for public safety."

The Economics of
Fire Protection Engineering

GERALD L. MAATMAN, P.E.

Any attempt to examine the engineering economics of fire protection should be done only within the context of risk management. As the term is applied to a business corporation, risk management requires the evaluation of all possible risks of loss, excluding inherent business risks. The first step in evaluating these potential losses involves a risk analysis—i.e., a study of the risks a company faces and their respective potential economic impacts on its profit-making ability and assets. In addition, where certain risks have life-hazard implications for employees and/or the public, they can sometimes take precedence over all other types of risk considerations.

Once the risks have been thoroughly evaluated, it is necessary to determine the most effective way to control these risks so that their impact in terms of cost is minimized.

Treatment of risk can take three forms:

- A risk can be *assumed* and its consequences accepted if it occurs.
- It can be *transferred* to some other risk bearer, such as an insurance company.
- A *protection* system can reduce or possibly eliminate the source of the risk.

In practice, owners handle risks through a combination of these techniques. They buy insurance to transfer some of their risks; they assume other risks; they take various protective steps to reduce the possibility of risks.

Business practices, as they relate to assumed risks, vary widely both by type of risk and size of organization. There is, however, a tendency for companies to assume an increasing amount of what they consider "normal" risks and to buy insurance only for risks of catastrophic nature. Thus, the benefits gained by reducing or eliminating assumed risks become increasingly important and are used to justify the cost of eliminating the risks through protection systems.

The economics of fire protection has its greatest application within the framework of risk reducing efforts and fire protection requirements. When there is a transfer of risk, the insurance carrier frequently imposes fire protection requirements on the company as a condition of obtaining insurance. It may also vary its insurance price depending on its assessment of existing property risk. When a company has assumed risks, it may voluntarily institute certain fire protection measures. Another source of fire protection requirements are building and fire prevention codes. These are imposed by government agencies mainly to protect adjoining property, tenants, and the public.

Questions to Be Answered

In any type of cost/effectiveness analysis, questions that must be answered include:

- What is the problem to be solved?
- What solutions are possible?
- Which solution will give the best result, taking into account relative costs and benefits?

In the cases of fire or explosion risks, identification of problem hazards is usually not too difficult. However, calculation of the statistical probability of the event and quantifying the resulting anticipated loss are often difficult and usually imprecise.

Fire and explosion losses can be divided into two subcategories: direct and indirect losses.

Direct losses include damage to the building, its machinery, equipment and contents. Although not usually considered in this context, the costs of personnel injuries and deaths resulting from fire or explosion are also part of direct loss. Depending on the building owner's degree of liability, these losses can exceed the cost of all other direct losses.

Indirect losses include loss of profits caused by the interruption of business, continuing expenses, permanent loss of customers and loss of skilled employees who take other jobs during their lay-off.

Determination of expected direct losses is usually made on a judgment basis. Taking an industrial plant as an example, an analysis is first made of the number and extent of separate fire areas in the plant. (A fire area is that portion of the total plant that would suffer damage if a fire occurred within that area and was allowed to burn itself out.) Boundaries between adjacent fire areas are usually defined by clear space, an adequately constructed fire wall or some other suitable means of preventing the spread of fire. Within a fire area, an analysis is made of damage that could result under three conditions: probable maximum loss (PML), maximum possible loss (MPL) and maximum foreseeable loss (MFL). These represent varying degrees of reliability of available fire protection measures.

PML is that loss expected if all available protection measures are operable and function properly at the time of the fire. These measures include automatic sprinklers and detection equipment, the prompt notification and response of the municipal fire department and any private fire brigade, and the automatic closing of fire doors.

MPL anticipates the failure of an important fire protection system component, such as a fire pump or sprinkler control valve. Under this condition an estimate is made of the loss that would occur if only the fire department and brigade facilities were available to fight the fire.

MFL presumes that the fire department will not be called or will not respond for some reason (adverse weather conditions, damage to communicating roads, etc.) and that the fire will progress until it burns itself out.

In the insurance field, these three types of analyses are generally made both in terms of direct and indirect losses. Underwriters establish their coverage limits, rates, and fire protection requirements based on results of these analyses.

It should be noted that these potential-loss analyses exclude the costs of injury and possible death of personnel. They also omit several indirect losses. They are really only guesses of the extent of damage that could occur under the three conditions. No data is available to predict the frequency of the MPL or MFL conditions occurring as opposed to the PML condition. Their use by a fire protection systems designer, risk manager or architect must be tempered accordingly.[1]

Estimates and Guesses

In designing necessary fire protection countermeasures, the engineer can only roughly estimate the total expected loss under the three assumed conditions of fire protection facilities reliability MPL, MFL and PML, none of which can be defined in terms of statistical probability. The designer can only place various types of fire hazards and occupancy types in a rough relative order of their probabilities of causing a fire.

These analytical shortcomings necessitate an approach by which the fire protection system designer must somewhat arbitrarily define a reasonable level of safety based on his judgment of the total values at risk, degree of life hazard, relative fire hazards presented by the occupancy, and the "norm" for the type of situation as determined by past fire experience and the codes or standards that cover it.

Using this "reasonable safety level" as a base, the fire protection system designer should increase protection where the total value at risk (direct or indirect) is abnormal, either in relation to the cost of significantly reducing the risk or to the total financial worth of the enterprise under consideration. He should decrease protection where risks do not warrant such expenditures.[2]

Once the designer has determined the level of fire protection needed, he must determine how to achieve

[1] The attempts to quantify the expected fire loss are less than scientific, but measuring the statistical probability of the fire or explosion ever occurring in a given type of occupancy or from a particular type of hazard is even less scientific. Despite nearly 100 years of organized record-keeping of fires, there is little knowledge of finite relationships between either given types of occupancies or individual fire hazards and their corresponding predictive frequency of fire. Insurance industry statistics, until quite recently, were gathered on a broad occupancy classification basis, and no organized attempts have been made to collect accurate data on the total existing exposure bases for various types of occupancies or hazards. Consequently, individual types of fire hazards and occupancy classes must be graded in relation to their respective potential to produce a fire.

[2] Although our fire protection engineering economics approach up to this point is something less than is desired, in the hands of an experienced engineer the intuitive judgments that must be made can still produce a reasonably effective allocation of fire protection resources and a rational solution to handling the company's property risks.

this level. Here is where the fire protection system designer can exercise a much greater degree of formal engineering economics methodology.

Essentially, he should develop two or more equivalent fire protection solutions to the problem, keeping in mind that each may have to satisfy varying needs (e.g., life-safety requirements, reductions in total direct and indirect losses, local code demands, insurance requirements). He then should conduct a conventional engineering economics analysis of each solution and choose the one that has the lowest total costs (first costs, operating costs, maintenance and depreciation factors, etc.).

This is not a simple analysis and set of calculations.

Constraints on Designers

Fire insurance requirements have exerted a strong influence on fire protection expenditures. Insurance rating practices, which offer reductions for certain features and penalty charges for others, directly affect the owner's costs and significantly influence the economic balance of alterntive fire protection solutions. (The presence or absence of complete automatic sprinkler protection is an excellent example of this factor.) In the same manner, insurance carriers' "highly protected risk" fire protection standards, which result in lower rates for property that meets the standards, has caused industry to build great reliability into its fire protection systems. This reliability takes the forms of secondary water supplies and sources of power, central station supervision of important fire protection system features, watchmen service, etc.

Many companies that have large concentrations of value in one fire area must install extensive fire protection systems in order to get insurance, regardless of price. Companies with this problem that lack the financial capacity to assume all their risks often need highly reliable fire protection systems regardless of the systems' reasonableness from a cost/effectiveness standpoint.

Although insurance cost considerations are important within the overall cost/benefit scheme, they do not constitute the whole analysis. As more companies assume increasing amounts of their total property risks, the effects of insurance cost obviously become less. This can work in favor of alternative fire protection solutions. Finally, an owner and his building designers should not assume that the insurance carrier's fire protection requirements are always fixed. In a significant number of cases, the use of a professional fire protection engineering consultant can produce less costly alternatives that are acceptable to the carrier.

Codes

The other principal outside constraints on fire protection system designers are applicable building and fire prevention codes. Code requirements are usually directed more toward life safety (both occupants and fire fighters) and protection of adjoining property than they are for the property in question. In addition, codes necessarily contain a generalized set of requirements applying to a wide variety of situations. Often a code requirement has no application in a specific situation and cannot be incorporated into any rational fire protection solution. Nevertheless, code approval must be obtained. A fire protection engineer often can develop alternative methods to meet the intent of a code and negotiate their acceptance. A protective system installed to eliminate a stair or a smoke tower in a high rise frees space for rental and is often a good code trade-off.

Fire protection standards are an area where the designer can achieve economies. For example, the sprinkler system design requirements of NFPA 13 are generalized for such things as pipe sizing and the number of heads per branch line. Strict adherence to these standards sometimes results in an overdesigned system from a hydraulic standpoint. A hydraulically designed system, as an alternative, can meet the fire protection needs of a property and, when properly documented, satisfy the requirements of insurance and code authorities.

To Prevent or to Control?

In developing a viable solution to a fire protection problem, there are usually two basic approaches available. One prevents fire from occurring; the other controls fire after it has started.

The prevention approach emphasizes the elimination of sources of ignition (e.g., provision of explosion-proof electrical equipment, bonding and grounding of flammable liquid dispensing equipment) and the reduction or elimination of flammable fuels (e.g., by substituting nonflammable solvents). Under exceptional circumstances, this method calls for elimination of oxygen from an area and substituting an inert gas.

The control approach can emphasize passive restraints (fire doors, fire walls, fire resistive floors) or active restraints (sprinklers, automatic fire detection, special application extinguishing systems).

Most often, a combination of the two methods is used. The passive restraints ensure against any unreliability in the active fire prevention measures.

The prevention approach generally is more economical in the long run. Very often, however, it is impractical from an operational standpoint, and full reliance must be placed on the control approach.

In the control approach, engineering economics usually favors active countermeasures because they result in lower insurance rates. Passive countermeasures, however, should not be overlooked. Building codes may require them. In some instances, an economical and effective solution can be obtained by combining a passive approach with "spot" active (and automatic) protection facilities for hazardous areas.

In factories, for example, the largest combined direct and indirect values at risk normally are in production areas. Consequently, the level of fire protection for such areas will be great. Innovations, such as high stockpiling in warehousing facilities, often produce a concentration of contents values that are large enough to dictate greater fire protection expenditures than normal.

In nonindustrial buildings, life-safety requirements frequently represent the most fruitful area for achieving economies in fire protection system design. For example, the judicious use of the horizontal-exit concept often permits elimination of at least one interior stairway that would normally be required. Aside from construction savings, the elimination of a stairway can produce more rentable space in the building. Economies can also be obtained by optimum design of such areas as exit sign layouts, fire alarm stations and HVAC system fire protection needs.

References

"The Economics of a Fire Protection Program," E. C. Burtner. 1965 Fall NFPA Conference.

"How Economics Guide Applied Research on Fire Protection Systems," W. M. Livingston. 1967 Annual Meeting, Society of Fire Protection Engineers.

"Engineered Sprinkler Protection," R. M. Patton. *Fire Journal,* January, 1966.

"Economics of Fire Prevention," F. J. Zeleny. *Building Research,* Jan.–Feb. 1965.

"Economics of Fire Protection," F. H. Gage. 1961 Fall NFPA Conference.

Fire Protection for Structural Components and Assemblies

JACK A. BONO, P.E.

The fundamental objective of structural fire protection is to confine a fire, to localize it until it can be extinguished. Structural members should be protected from collapse, and walls, partitions, floors and roofs should be designed to resist flame and heat penetration. To achieve this objective, consideration must be given to how long a building may endure a fire and what intensity of burning can be anticipated.

Shoub, in a paper on fire endurance testing [1], describes efforts made soon after 1800 to use noncombustible construction to provide fire resistance; among the materials used were masonry exterior, interior framework of cast iron columns and iron girders and floor beams. Unfortunately, this construction distorted and collapsed in fire—and we still need to learn more; the destruction of Chicago's McCormick Place in January 1967 stands as mute testimony to the fact that the terms noncombustible and fire resistant are not synonymous.

Early fire testing was directed toward evaluation of individual materials rather than assemblies but floors were tested as early as 1890, in Denver. The floors were subjected to load, shock, fire, and water. Tests were made for periods as long as twenty-four hours but at a lower temperature than now employed in tests. Some tests used gas fuel; others, such as a comprehensive series conducted in 1896 by the New York Building Department, used wood fuel. Maximum temperatures in the New York tests ranged from 1975° to 2575° F. Columbia University made major contributions by its fire tests of floors [2], reported in 1912. In 1905, the

predecessor to ASTM Committee E-5 was formed and a floor test standard was adopted in 1908.

There were few tests of partitions until more recent times, although Columbia University was active in this area as well as in floors. Records of tests of columns parallel the early floor tests. The most extensive series of tests of columns was conducted in 1917 and 1918 at Underwriters Laboratories in Chicago, with the cooperation of the Associated Factory Mutual Fire Insurance Company, the National Board of Fire Underwriters, and the Bureau of Standards [3].

Development of Test Standards

The most notable development from this early history was a conference in 1916–17 leading to the adoption of a test standard not substantially different from the current standard published by ASTM as ASTM E-119, by the National Fire Protection Assn. as NFPA 251, and Underwriters Laboratories as UL263.

The early history of fire testing teaches that the easiest testing is conducted on individual materials, but the better evaluations involve tests of assemblies.

Fire Severity and Fire Load

The severity of fire exposure included in the test standards is based on a standard time-temperature curve (see Figure 6.1). It is a composite of a considerable number of performances and calculated curves studied by the 1916–17 conference. It is characterized by a

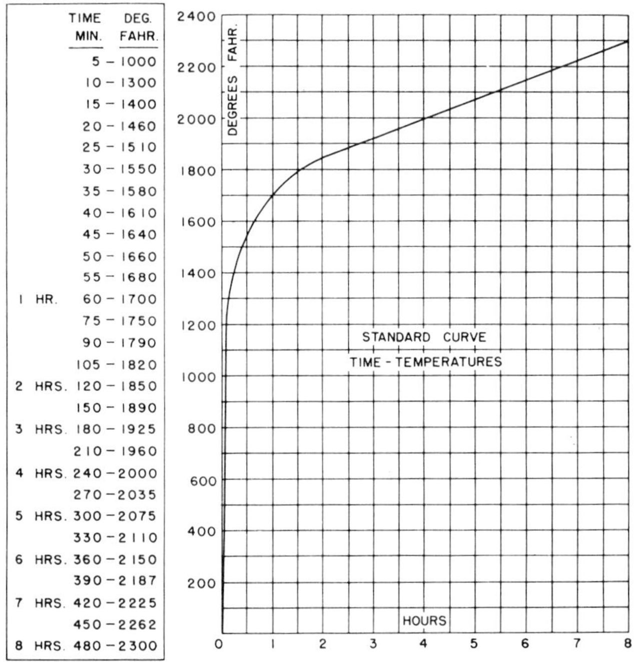

Figure 6.1.

rapid rise in temperature during the first ten minutes of the test.

Despite the use of this curve in hundreds of tests, its applicability has always been debatable. It would seem that the fire severity should be related to the combustible material available in a building, including contents as well as elements of the construction itself. This combustible quantity is called the fire load. The relationship between fire severity and fire load was explored in a series of burn-out tests [4] performed in "fireproof" structures. Various concentrations of combustibles having a calorific value in the range of wood and paper (7,000–8,000 Btu/lb), assembled to represent building occupancies, were used. The results are summarized in Table 6.1. A survey [4] of combustible contents or fire load of buildings developed approximations for various occupancies as shown in Table 6.2.

Knowledge of the fire load coupled with the antici-

Table 6.1. Fire Load vs. Fire Severity

Average Wt. Combustibles $(lb/ft^2$ of floor area)	Fire Severity Hrs.
5	½
7½	¾
10	1
15	1½
20	2
30	3
40	4½
50	6
60	7½

Table 6.2. Survey of Combustible Contents of Buildings

Occupancy	Average Combustible Contents-lbs/ft^2
Apartments and Residences	8.8
Offices (only)	7.9
Office and Reception Room	6.6
Office and Light Files	10.9
Office and Heavy Files	42.9
School Buildings	Less than 10
Hospitals	5.0
Warehouse-Printing Dept.	166.8
Warehouse-Department Store	14.3

pated fire severity appears to be the basis for code requirements for fire-resistance ratings of building assemblies and individual members. These two factors influence building code provisions, but there are other considerations.

Data on combustible loading was published in 1942. A comprehensive study has not been conducted in the United States since. The nature of occupancies has changed. Metal furniture is often used. Interior finishes for certain occupancies are required to be of low combustibility. On the other hand, materials of ordinary calorific value (8,000 Btu/lb) are occasionally replaced with materials having values of 14,000 Btu/lb. Storage occupancies in particular have changed in character. High stockpiling of materials (see Figure 6.2) results in fire loadings far in excess of those with which we have been familiar. Pilings of forty feet or higher are common. This permits fire development (see Figure 6.3), often despite automatic sprinkler protection, which exposes structural members to temperatures that can bring about collapse.

Within a particular occupancy, the combustible content may vary from one area to another, and knowledge of the content is not enough to anticipate severity.

The required fire resistance for structures is influenced by the location of the building with respect to fire

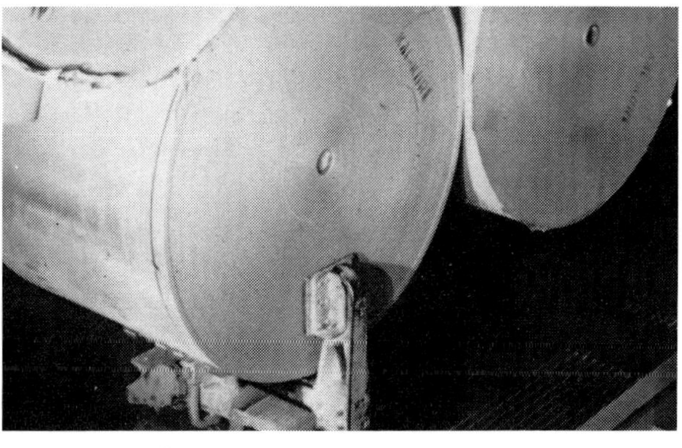

Figure 6.2. High piled paper stock.

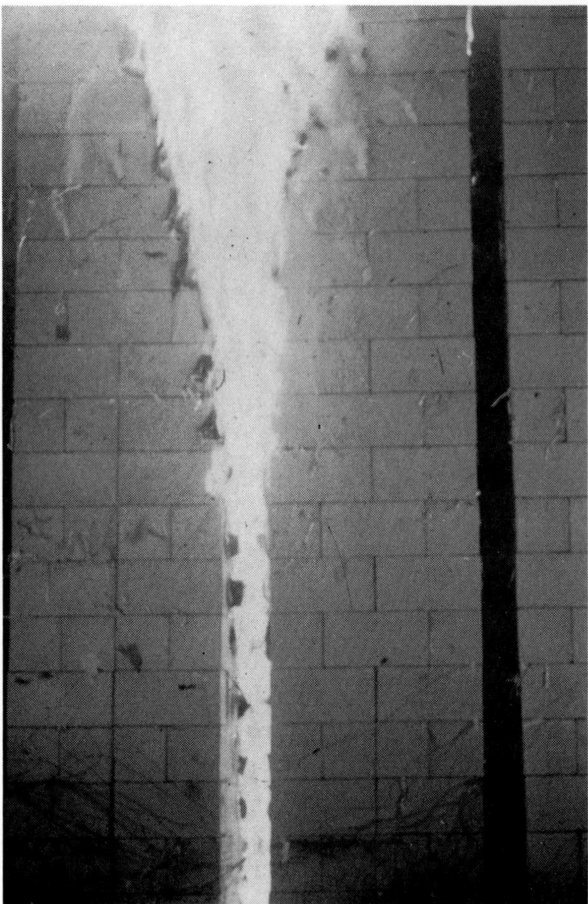

Figure 6.3. Fire in piled stock, 30 seconds after ignition.

zones. A congested and industrialized area normally requires greater fire resistance to avoid conflagration.

Danger from exterior fire exposure also is a factor in code requirements for structural fire protection. For example, in one building code, the fire resistance rating for an exterior, nonbearing wall is specified at zero, one and a half, and two hours for a building separated from another by thirty feet, eleven feet, and six feet respectively [5].

Height and area requirements dictate the degree of fire resistance required for structural assemblies and walls. In most cities, the height and area of buildings of other than fire-resistive construction are limited. This helps the fire department handle fires in such structures, particularly in upper stories.

The nature of the occupancy is another factor in the required structural fire protection, beyond the consideration of the fire loading. The number and concentration of occupants and whether such occupants are able-bodied and not restrained or nonambulatory and restrained affect requirements. There are other considerations, too, such as whether sprinkler

protection is required, the availability of fire departments and water supplies and, of significance, the cost of materials and erection and the cost of insurance protection.

It is apparent, then, that the nice, simple relationship between fire loading and structural protection has been removed from its understandable but isolated status and placed in context with other factors, some of which are subjective. This is why unanimity of opinion and code requirements does not exist with respect to required structural fire protection. Nevertheless, it was concluded in a paper delivered in 1965 by a member of the steel industry that "the results of this study have indicated that to a great degree there is far better uniformity in codes than ever before. Although inconsistencies exist, many have been removed from codes and work is in progress to minimize those that remain."

Standard Test

Contributing to the increased uniformity of requirements is the fact that there is a standard test for use in classifying the fire resistance of floor and roof assemblies, walls and partitions, and individual structural members, such as columns and beams. This standard, officially known as Standard Methods of Fire Tests of Building Construction and Materials, is referenced in each of the four model codes published in the United States (see Chapter 2) and is followed in principle in other countries.

The classification or rating, expressed in hours, is intended to register performance during the period of exposure and is not to be construed as having determined suitability for use after fire exposure. It is important to recognize that an assembly classified as two hours, for example, may collapse at two hours and one minute. Therefore, if a safety factor is desired, it should be employed in the code provision.

How Standard Test is Conducted

For test purposes, floor and roof assemblies are constructed in a frame having inside dimensions of approximately 18 X 14 ft. (see Figure 6.4). The assemblies dry to a uniform moisture condition and then are placed over a floor furnace (see Figure 6.5), where they are subjected to fire from below.

These assemblies and the structural members that form a part of the assembly are required to be restrained in a manner simulating restraint in actual conditions. Unfortunately, there is no clear understanding of the degree of restraint to be anticipated in actual construction, since it varies, depending on whether

Figure 6.4. Construction of floor test sample.

the assembly is in an interior bay or elsewhere in the building. Restraint is a major influence in the structural performance during the test. The assembly is loaded to create maximum design stress in structural elements. This is accomplished by placing water-weighted pans on the top surface or by using hydraulic jacks.

Requirements of Standard Test

The assembly must withstand the full fire exposure for the classification period without collapse. In addition, no openings or cracks shall develop in the assembly that will permit hot gases to penetrate and ignite combustibles on the top surface.

Finally, the assembly must resist transmission of heat so that the temperature on the top surface or side unexposed to fire does not exceed an average rise of 250° F. or 325° F. at any point. The purpose of this last criterion is to prevent the ignition of combustibles on the floor above the fire. These are the criteria to be met if the assembly is to do its job of confining a fire.

At one time, the test standard specified that after the fire exposure, a hose stream was to be directed on the underside of the assembly and that no collapse or penetrations should occur as a result. The hose-stream

Figure 6.5. Top view of floor test furnace.

test was eliminated from the standard for floor and roof assemblies, though some codes or enforcing authorities still require it to be conducted. For that reason, the hose-stream test is employed on an option-basis for floor and roof assemblies.

Test for Walls and Partitions

A wall or partition is also constructed into a test frame (see Figure 6.6). This frame is placed in front of a vertical panel furnace, where the sample is exposed to fire. The size of the sample must be not less than ten square feet with neither the width nor height being less than nine feet. Load-bearing walls are again subjected to applied loads.

The criteria or conditions of acceptance are similar to those for floor assemblies, with respect to resistance to collapse, development of openings and heat transmission. However, the hose-stream test is mandatory for assemblies having a fire resistance rating of one hour or more. The hose-stream test may be conducted on the assembly after the fire exposure or on a duplicate specimen that has been subjected to a fire-exposure test

Figure 6.6. Construction of wall test sample.

for a period equal to one-half of that indicated as the resistance period, but not for more than one hour.

The hose-stream test causes substantially more failures of walls and partitions in comparison with floor and roof assemblies. The hose stream is intended to create an impact strong enough to assure that the assembly has some structural integrity to resist falling timber or stock during a fire.

Test for Building Columns

Building columns are tested in a furnace with a fire that creates a uniform temperature condition around the column, following the standard curve. The height of the sample shall be not less than nine feet.

Many early tests of columns were conducted under load in order to develop as nearly as practicable the working stresses contemplated by the design. In the case of steel columns, however, it has been fairly well established that an average temperature of 1000° F. in the column means structural failure is imminent. This temperature is slightly conservative, but even recent tests tend to bear out this temperature/structural failure relationship.

Current Trends in Test Standard

Revisions of this test standard have been made through the years, mostly to introduce greater accuracy and uniformity in the methods. Specifications for controlling the moisture, particularly in concrete samples, have been placed in the standard.

The moisture content is a factor in the ability of concrete to resist heat transmission, and the revisions have improved the standard. In addition, the instrumentation for this test for measuring temperature and deflection has become more sophisticated.

The fundamental concepts and criteria, however, have not been changed. One significant change was brought about by the recognition that the manner in which a structural member or assembly is restrained, that is, is kept from expanding when subjected to heat from fire, influences its structural performance. A steel beam which is unrestrained and simply supported so that it is permitted to expand in a fire test, will lose its strength and collapse when the temperature of the steel reaches the vicinity of 1000° F., if it is loaded to its maximum design load. With optimum restraint against expansion, the same beam will resist collapse until the temperature of the steel reaches substantially higher values.

Ratings of floor-ceiling and roof-ceiling assemblies and structural elements in those assemblies are now designated as restrained or unrestrained, depending on the manner of test and the criteria applied in judging the results.

How Structural Fire Protection is Achieved

Generally speaking, structural fire protection is achieved by using materials that have inherent resistance to fire or by using forms of fireproofing to shield structural members and assemblies.

The most common material with inherent fire resistance is concrete. This is not to say that fire resistance is present in all shapes and designs of concrete, because the degree of fire resistance of concrete is influenced by certain factors.

The two most significant factors are thickness and aggregate type. Sufficient tests [6] have been conducted to establish a relationship of these characteristics with fire endurance. The fire endurance depicted by graphs in Figures 6.7 and 6.8 is based on maximum permissible temperature being reached rather than structural failure.

Effect of Concrete Cover on Steel

In addition to concrete thickness and type of aggregate, some influence is exerted by the thickness of concrete cover for the reinforcing steel. Since the steel is used to provide tensile strength and since the tensile strength is reduced with temperature, the concrete cover is required to shield the steel from the heat. This is particularly true for concrete sections used as simple span construction. If the concrete construction incorporates continuity or is framed in a manner that provides restraint to expansion caused by fire, the effect of concrete cover is diminished.

In tests [7] conducted by the Portland Cement Association, a number of prestressed double-tee floor sections were subjected to fire exposure, with each section permitted to expand a different amount before being restrained. The sections were held against rota-

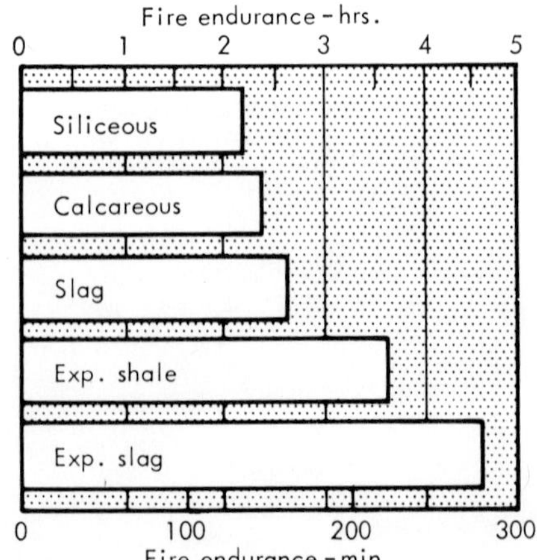

Figure 6.7.

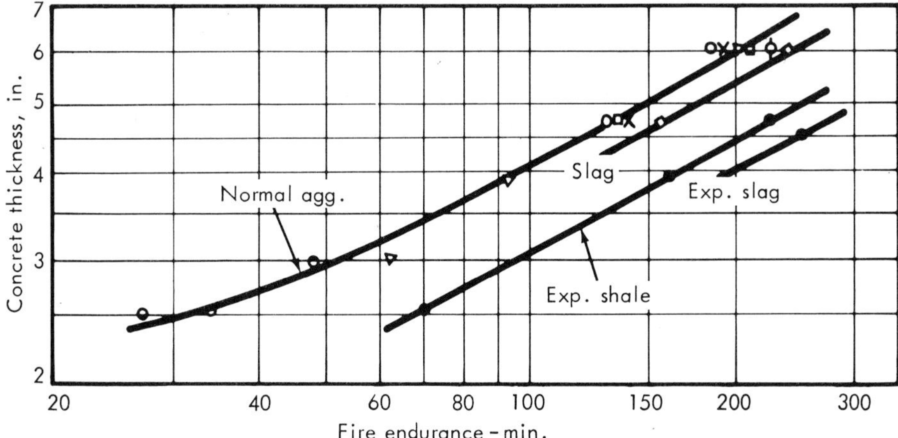

Figure 6.8.

tion in all tests. A summary of the results is shown in Table 6.3. With various intensities of restraint, achieved by different permissible expansions, the fire endurance exposure was four hours and ten minutes. Even at that time, the maximum deflection of any sample was 2.25 inches. The interesting point about these results is that the fire resistance of these sections, if tested while simply supported, would be approximately one hour. In other words, the increase in fire endurance as a result of restraint was at least 400 percent. The restraining forces place the concrete under compression so that it can resist the positive moment action of the imposed dead and live loads.

Restraint of concrete sections does not always act to the best interests of this form of construction. In recent years, many tests have been made of prestressed concrete slabs placed in a heavy test frame and grouted to fill any space between the ends of the slabs and the frame. On occasion, the result has been a violent disruption of the concrete, exposing reinforcing tendons and creating openings in the assembly as shown in Figure 6.9. There is some debate whether this development will occur in actual construction, since the degree of restraint in a building may be less than in the test frame.

A method for increasing fire resistance for concrete construction was tested [8] in 1964 by Underwriters Laboratories, Inc. The construction consisted of a re-inforced-concrete pan-and-joist assembly with 3/4-inch

Figure 6.9. Prestressed concrete after explosive failure.

thick cementitious mixture fireproofing applied to the bottom of the pan sections only (see Figure 6.10).

The thickness of the pan section was 2 3/8-inches and the estimated fire endurance rating if no fireproofing had been used was approximately thirty minutes. The actual fire resistance of the assembly was one hour and thirty-nine minutes. Other data established that a 2 1/2-inch thick concrete pan section with 7/8-inch thickness of fireproofing would have a fire resistance rating of two hours.

A very common form of floor and roof assembly utilizes metal lath or a metal deck floor on top of which

Table 6.3. Results of Restrained Prestressed Sections

Sample No.	Allowed Expansion (In.)	Total Deflection At Midspan (In.)	Duration of Test Hr-Min.
5-19	0.121	−0.15	4 hr. 10 Min.
5-11	0.289	0.25	"
5-10	0.552	0.35	"
5-12	0.750	1.20	"
5-13	1.000	1.74	"
5-17	1.210	2.25	"

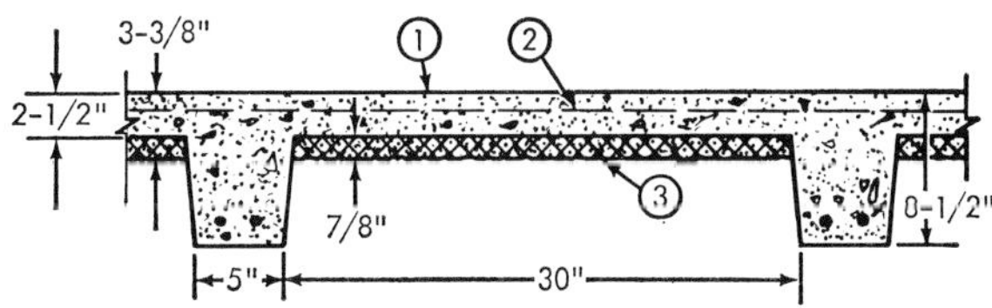

Figure 6.10. Cross-section of pan and joint construction showing cementitious-mixture fireproofing.

concrete is poured. The structural members for this
assembly consist of steel beams or steel joists. The
problem is to protect the metal members from heat,
because steel loses its structural strength at fire temperatures.

Restraint Influences Steel Construction

As with concrete, structural performance of a steel
beam is influenced by restraint. In a series of tests [9]
similar to those conducted for prestressed members
as described earlier in this chapter, restrained beams
showed an improvement in fire resistance of approximately 25 percent compared with unrestrained beams.

In these tests, the bottom flange of the steel beam
developed a buckling failure approximately eighteen
inches from the ends. Also, the degree of restraint did
not seem to be a major factor in the structural performance. The 25 percent improvement occurred
under a variety of restraint conditions.

Applied Fireproofing

A common way of protecting steel from the heat
of the fire is by directly applied fireproofing. One type
consists of a cementitious mixture of lightweight aggregate, vermiculite or perlite, and gypsum. Another is
sprayed fiber consisting of mineral wool or asbestos
fibers mixed with clay and a few minor ingredients.
Both these fireproofing materials may be applied by
a nozzle.

Another form of fireproofing is a gypsum plaster
applied over a caged beam. All fireproofing materials
containing gypsum depend on the inherent characteristic of the gypsum to calcine upon heat exposure and
release moisture. Vaporization of the moisture uses the
heat from the fire source and, therefore, delays heat
transmission to the steel.

Sprayed-on fireproofing materials are not only
applied to structural members but also are applied to
metal deck. When a header is used in the assembly, a
somewhat thicker application may be required under
these elements to resist heat transmission. Sprayed-on
materials provide a convenient form of construction.
Some of these, however, are soft and more easily
damaged than other types of fireproofing. Also, mechanical contractors find it easy to remove the fireproofing
in order to accommodate piping, ducts and other equipment.

In one city, the building official required that the
material be made more resistant to casual damage and
removal, and the sprayed-fiber manufacturers developed
an overspray for this purpose. A committee of the
American Society for Testing and Materials is currently
considering the possible development of a test for

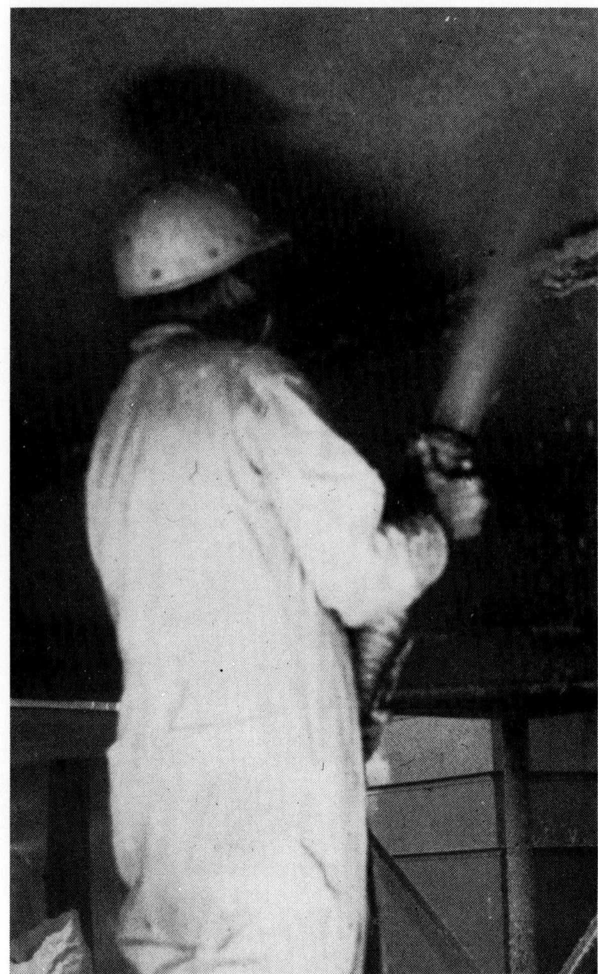

Applying sprayed fireproofing.

measuring the damageability of sprayed-on types of
fireproofing.

Membrane Fireproofing

Since 1960, the membrane type of fireproofing
(see Figure 6.11) has become popular. Early membranes
consisted of gypsum plaster and lath. More recently
membranes have included acoustical tiles and panels.

There are two types of acoustical membranes, one
in which the tee sections of the suspension system and
the splines fit into slits in the edges of the tile, and
another consisting of a lay-in system in which acoustical panels are supported by a grid. With the lay-in
system, hold-down clips are required to resist lifting
of the acoustical panels in case of pressure pulses created by slight explosions, closing of doors, etc.

Acoustical tile used in these systems is not ordinary
mineral tile. It has special properties, particularly with
respect to shrinkage resistance, in order to prevent tile
from falling out of the suspension system during a

Figure 6.11. Membrane ceiling protection.

fire. The material is specially batched with clays, binders and other ingredients.

The membrane-protected assemblies are often installed and tested with lighting fixtures. On the basis of acceptable performance in these tests, maximum areas of lighting fixtures are permitted. Indiscriminate substitution of lighting fixtures in assemblies tested without fixtures can lead to premature failure by heat transmission or fallout of the acoustical panels. Duct openings are also provided in these membrane ceilings, and such openings are normally dampered. When the membrane ceiling is used, a plenum is formed between the ceiling and the underside of the floor or roof assembly. Heat that penetrates the membrane collects in the plenum, raising the temperature around structural members. Some of this heat is absorbed into the floor or roof, and the rest is retained in the plenum.

Here again, an indiscriminate substitution in a tested floor or roof of materials having different thermal capacities and insulating characteristics may cause the heat to be retained in the plenum, resulting in premature failures. Tests in which the concrete roof of a membrane type of assembly has been replaced by an insulating board have revealed a reduction in fire resistance from two hours to forty-five minutes. This is why fire-protection engineers agree that there is no such thing as a two-hour ceiling or one-hour ceiling. The fire resistance is provided by the assembly.

Protection of Walls

Fire-resistant walls are often constructed by facing both sides of an interior structural member, such as wood or steel studs, with a fireproofing material such as gypsum plaster or gypsum wallboard with a steel facing. Predominantly, wall constructions are homogeneous in the sense that the fire resistance is the same when fire occurs on either side.

A fairly inexpensive form of fire-resistant wall construction is lightweight concrete block or masonry unit. Up to four-hour fire resistance can be obtained with the nominal 8 X 16-inch block with a plaster face or with the core openings filled with an insulating material. Fire resistance of concrete blocks varies with the type of aggregate, the equivalent thickness and the ratio of aggregate to cement.

Protection of Columns

Fire resistance of columns is achieved in much the same way as protection for steel beams—either by directly applied fireproofing, by plaster on caged columns, or by gypsum wallboard. Prefabricated columns made of concrete filled steel pipes with a steel structural interior member have earned fire resistance ratings up to four hours.

Variations in the prefabricated column include filling with a lightweight aggregate material and use of rectangular or square steel sections.

Openings

One of the major problems of structural fire protection is to avoid unprotected penetrations in the fire resistant assembly. Openings are required for electrical outlets, telephone service, air ducts and even expansion joints.

Some experiments have been conducted using different methods to close such openings, but so far, no major effort has been expended in attempting to solve this problem. Evidence in the McCormick Place fire suggests that the fire was conveyed from the third floor to the first floor via an expansion joint through which molten aluminum seeped.

Large Assembly Testing

Some testing of larger size members and assemblies is being accomplished. In a test [10] of a large-size concrete wall in Japan, cracks caused by thermal expansion raised a serious question of the capability of the wall beyond one hour. A wall of similar construction was tested in the usual size of test sample and achieved a two-hour fire resistance rating.

Fire walls are self-supporting and should be designed to maintain structural integrity in case of complete collapse of the structure on either side. To withstand heat expansion effects, they are commonly made thicker than would be indicated by normal fire resistant rating requirements, and, if of considerable height or length, they should be buttressed by cross walls or pilasters.

Fire walls must extend through and above combustible roofs to prevent the spread of fire. Fire walls ordinarily need not be extended through fire-resistive roofs, decks.

Publications of Fire Resistance Ratings

Much information has been accumulated on fire resistance ratings of building assemblies and structural members. This information is published by organizations responsible for the four model codes in the United States.

The tabulation of fire resistance ratings may be a part of the code, as in the case of the Southern Standard Building Code, or may be contained in a specific document, such as the Fire Resistance Ratings published by the National Board of Fire Underwriters (now the American Insurance Association). Underwriters Laboratories, Inc., publishes an annual Building Materials List. The 1967 list contains 530 illustrations of columns, walls, and partitions and floor and ceiling assemblies having fire resistance ratings of up to four hours.

An excellent compilation [11] of concrete assemblies is published by the National Research Council of Canada, which also has a similar compilation [12] of protected steel columns and beams.

There are other sources for information on fire resistance ratings, many of them contained in individual building departments.

This information can be the basis for making interpretations of assemblies that have not been subjected to the fire resistance test. However, these interpretations should be made by knowledgeable fire protection engineers totally familiar with factors that influence performance of structural members in fires and who have access to data on this performance.

A project at Underwriters Laboratories involves the proper coding of information for programing into a computer. It is anticipated that the computer can be utilized to provide instant data concerning the effect of a large number of variables in construction. In this instance, as in other scientific ventures, modern technology is being used to assure the greater safety of people and to reduce losses from fire.

References

1. Shoub, H., "Early History of Fire Endurance Testing in the United States," Special Technical Publication No. 301, American Society for Testing and Materials (1961).

2. Woolson, I. H. and Miller, R. P., "Fire Tests of Floors in the United States," Proceedings, Sixth Congress, Intl. Assoc. for Testing Matls., Second Section, SSVII (1912).

3. "Fire Tests of Building Columns," Underwriters Laboratories, Inc., 1919.

4. "Building Materials and Structures," Report BM592, National Bureau of Standards (Oct. 7, 1942).

5. "BOCA Basic Building Code," Fourth Edition, Table 5, Building Officials Conference of America (1965).

6. Benjamin, I. A., "Fire Resistance of Reinforced Concrete," American Concrete Institute Publication SP-5 (1961).

7.   Selvaggio, S. L. and Carlson, C. C., "Effect of Restraint on Fire Resistance of Prestressed Concrete," STP No. 344, 1–25, American Society for Testing and Materials (1963).

8.   "A Concrete Pan and Joist Reinforced-Concrete Slab Protected with Cementitious Mixture," Underwriters Laboratories, Inc., Report (1964).

9.   Bletzacker, R. W., "Effect of Structural Restraint on the Fire Resistance of Protected Steel Beam Floor and Roof Assemblies," Final Report EES 246/266 (1966).

10.   Kawagoe, Kunio, "Damage of Structures in Full-Size Fires," Report of the Building Research Institute No. 29 (Japan) (March 1959).

11.   Galbreath, M., "Fire Endurance of Concrete Assemblies," Technical Paper No. 235, Division of Building Research, National Research Council of Canada, (Nov. 1966).

12.   Galbreath, M. and Stanzak, W., "Fire Endurance of Protected Steel Columns and Beams," Technical Paper No. 194, Division of Building Research, National Research Council of Canada (April 1965).

Part II

Water Supply

AND

Water Related Fire Protection Systems

Water Supply Systems Design

ROLF JENSEN, P.E.

The basic input design criteria for fire protection water supply systems are:

- To evaluate the quantity and pressure needed at the point of delivery.
- To establish design goals based on present and projected needs and expressed as system demands, building locations and economic constraints.
- To plan design goals that will meet or be compatible with water company and insurance company requirements.
- To select and locate storage and pumping facilities when needed.
- To select and arrange piping to transport water from this supply source (storage or otherwise) to the point of delivery based on the flow and pressure requirements and the pipe laying conditions.
- To position or locate valves, controls and supervisory devices to ensure availability when needed (reliability).

The quantity of water required is generally determined by the design demands of a protective system using water. NFPA Standards 11, 11A, 13, 14, 15 and 16 each contain design guides for water needed (flow rates and storage quantities). NFPA standards that deal with specific occupancy conditions also contain design guides for the rate and quantity of water needed. Among these are NFPA 231, 231-B, 231-C, and 409.

In many jurisdictions the building, mechanical, or fire prevention codes contain water supply design requirements. Sometimes OSHA requirements prevail, quirements. Sometimes OSHA requirements prevail, and infrequently other standards (i.e., federal), codes, or guides apply.

Most of the time, the type and amount of insurance carried affects the water demand. When coverages are by superior-risk insurance companies, such as Factory Insurance Association (FIA), one of the member companies of the Factory Mutual Insurance System (FM), Kemper Insurance or similar insurers, the amount and rate of water delivery and number of supply sources will usually be based on the insurance "requirements."

As insurable values and hazards rise, it is common for the insurance company to require a highly reliable single source, dual sources, or occasionally more. These input design criteria must be established by the designer before design proceeds. They are not rigidly established and are often negotiable.

Water company design constraints are both economic and technical. Connection charges and metering requirements must always be considered. Often the water company will want full-flow turbine meters or detector checks, which are expensive "add-ons." When dual-source systems (city water and reservoir) are designed, water companies often require double-check valves or backflow preventers.

If nonpotable secondary supplies are used, health authorities will usually ask for backflow preventers, air gaps or automatic syphon breaks. In each of these cases, the high friction loss characteristics of metering or flow direction control devices conflict with the need for high flow/low friction-loss fire protection water supplies.

In most projects, adequate and economic design follows directly from the established water demand, storage and source criteria. However, when a building or plant is to be constructed in stages over an extended period (five to twenty-five years or more), it is often economically desirable to establish an orderly development program to assure that system capability grows with need and that capital expenditures are made when needed.

A long-range design plan to assure careful location of water sources and mains compatible with planned building locations may save considerable future reconstruction cost. This will become particularly important when the supply system serves multiple functions—fire, domestic, air conditioning, and process uses.

While many code and insurance authorities view such systems dimly, reliability can be assured with good design, planning, and controlled use. Substantail cost savings often accrue in combined systems, but the engineer must be prepared to prove his case, both to the enforcing authority and to his client, who must invest engineering time to save the capital expenditure.

Water Supply Sources

Selection of the water supply source or sources is the next step in design planning. Potential sources are city supply mains, lakes, rivers, on-grade and buried reservoirs, wells, gravity tanks and pressure tanks. Often the type of water supply selected determines its location in the system. Basic to the selection is the fact that water is moved by pressure differentials (energy) from source to point of demand.

An economic balance must be achieved between the cost of pumping when needed and the cost of piping. Since some of the piping cost may be in the demand system (sprinkler or other), the engineer must not lose sight of the piping in this system as a factor in achieving good economic design. He should be cautious of design conditions that introduce operating pressures above 175 psi, since such pressures require use of expensive, extra-heavy fittings.

When available, a city water supply is usually the least expensive and most reliable supply source. A current dependable flow test that records the capability of an available city supply is a necessary first step in evaluating the city supply source. Pumps can be used to improve a city supply when it is deficient in pressure, but they will not improve a city supply that is deficient in volume.

Lakes and rivers, when available, should be evaluated for continuity of supply (i.e., lowest mean level available for assured pump intake) and for water quality.

Beware of water supplies that are potentially corrosive, very hard or contain debris. Use of such supplies always makes pumping necessary. They are usually nonpotable and cannot serve as a second source to a potable city source. Many times a decorative lake or a swimming pool can serve a dual purpose as a fire protection supply.

Reservoirs

Reservoirs, once sized, are also developed by pumping. Available reservoir variations include buried steel or concrete, on-grade steel, or concrete and on-grade plastic-lined earth-formed. Reservoir reliability is enhanced by dividing, and sometimes this will serve to provide the equivalent of two sources.

Painting and cleaning must be considered in evaluating operating costs. Use of "Core-Ten" steel is usually a good investment to avoid exterior painting of above-ground steel tanks. Reservoirs must be heated when exposed to freezing temperatures. This increases initial and operating costs.

When concrete reservoirs are located in buildings, cost savings sometimes can be found by using swimming pools or by planning foundation walls to do double-duty.

Wells are sometimes developed as the water supply for a vertical fire pump. Excellent guidelines are contained in NFPA 20, Standard for the Installation of Fire Pumps.

Gravity and pressure tanks are very reliable. They are practically limited to an installed height of about 150 feet (65 psi) above grade or 50 feet (22 psi) above a building. Thus, they are not used when high pressure is needed. Gravity tanks must be regularly cleaned and painted and must be heated when exposed to freezing temperatures—all of which affects initial and operational costs. Pressure tanks are limited in quantity (normally 5,000 gallons) and require air pressure for operation. They have the same cleaning, painting and heating needs as gravity tanks.

Piping Layout

Having selected and located the supply, the designer must next proceed to piping layout. Primary goals of reliability and economics often dictate the use of looped systems with sectional control valves located so as to ensure that the dual path of a flow common to a single or compound loop is actually available. This latter point appears to be an oversimplification, but it is, unfortunately, poorly analyzed in many cases.

Control valves must be placed so that they assure two real paths of flow from source to points of use—and no more. Hydrants should be located on a spacing of 250 to 300 feet and closer when the mains are long

An example of how water supply demand is established can be taken from NFPA 13. Figures A and B (below) show tables and curves used for the selection of system flow rate, duration and minimum pressure.

As a general rule, the values in Figure A can be used directly in establishing a sprinkler system water demand requirement. The values in Figure B, which establish design criteria for hydraulically calculated sprinkler systems, should be multiplied by 1.15 because that much excess flow normally occurs as a result of hydraulic imbalance.

In both cases, pressure needs vary with piping friction loss and elevation changes. They can be estimated or calculated precisely as discussed elsewhere in this article.

Figure A. A Guide to Water Supply Requirements for Pipe Schedule Sprinkler Systems

Occupancy Classification	Residual Pressure Required (See Note 1)	Acceptable Flow at Base of Riser (See Note 2)	Duration in Minutes (See Note 4)
Light Hazard	15 psi	500–750 gpm (See Note 3)	30–60
Ordinary Hazard (Group 1)	15 psi or higher	700–1000 gpm	60–90
Ordinary Hazard (Group 2)	15 psi or higher	850–1500 gpm	60–90
Ordinary Hazard (Group 3)	Pressure and flow requirements for sprinklers and hose streams to be determined by authority having jurisdiction.		60–120
Warehouses	Pressure and flow requirements for sprinklers and hose streams to be determined by authority having jurisdiction. Also see NFPA 231 and NFPA 231C.		
High-Rise Buildings	Pressure and flow requirements for sprinklers and hose streams to be determined by authority having jurisdiction.		
Extra Hazard	Pressure and flow requirements for sprinklers and hose streams to be determined by authority having jurisdiction.		

Notes:
1.   The pressure required at the base of the sprinkler riser(s) is defined as the residual pressure required at the elevation of the highest sprinkler plus the pressure required to reach this elevation.
2.   The lower figure is the minimum flow including hose streams ordinarily acceptable for pipe schedule sprinkler systems. The higher flow should normally suffice for all cases under each group.
3.   The requirement may be reduced to 250 gpm if building area is limited by size or compartmentation or if building (including roof) is noncombustible construction.
4.   The lower duration figure is ordinarily acceptable where remote station water flow alarm service or equivalent is provided. The higher duration figure should normally suffice for all cases under each group.

Figure B. A Guide for Determining Density, Area of Sprinkler Operation and Water Supply Requirements for Hydraulically Designed Sprinkler Systems.

| Hazard Classification | Minimum Water Supplies | | Duration in Minutes |
	Sprinklers GPM	Combined Inside & Outside Hose—GPM	
Light	150	100	30
Ord.—Gp. 1	400	250	*60–90
Ord.—Gp. 2	600	250	*60–90
Ord.—Gp. 3	750	500	*60–120

*Notes:
1.   The lower duration figure is ordinarily acceptable where remote station water flow alarm service or equivalent is provided.
2.   For dry systems increase area of sprinkler operation by 30 percent.
3.   For combustible construction with wet or dry systems the minimum area of application shall be 3,000 sq. ft.
4.   For hazard classifications other than those indicated see appropriate NFPA Standards for design criteria.
5.   Calculations shall be based upon the area of sprinkler operation selected from the graph at right or upon the area of the largest room being considered, whichever is smaller. Such rooms shall be enclosed by construction having a fire resistance rating at least equal to the water supply duration indicated in the graph and wall openings shall be protected in an approved manner. For areas of sprinkler operation less than 1500 square feet, the density for 1500 square feet shall be used.

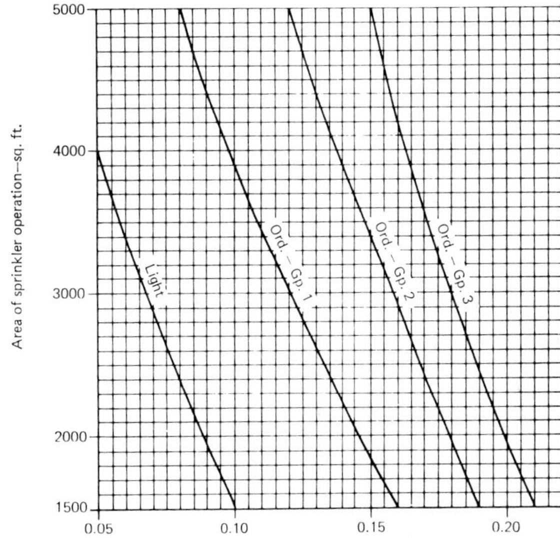

distances laterally from buildings. Hydrant location in private systems is based on the actual distance a hose must be laid to reach a fire. This is different from hydrants located for use by a fire department.

Piping is selected on the basis of internal pressure of the system and laying conditions, including type of soil, soil corrosiveness, trench conditions, unusual external loads and other criteria. Pipe size and type of material or lining also is based on expected flows and allowable pressure losses. An excellent list of pipe design standards can be found in NFPA 24, Standard for Outside Protection.

Considering the amount of time, effort and money expended on a good water supply and piping system, it is unfortunate that one small but important factor is often overlooked: closed valves account for 35 percent of the failures of fire protection systems. These are minimized by providing valve supervision, by use of easily observable indicating valves (PIV's) or by use of valves that cannot be left closed unless attended. These special valves became available in 1973.

How to Evaluate Supply Source

The evaluation of a supply source involves basic hydraulics, mostly friction-loss calculations. Commonly, fire protection friction-loss calculations use an empirical form of the Hazen-Williams formula in place of the more cumbersome Fanning or Darcy-Weisbach formula. Sufficient accuracy is achieved for most cases. The Hazen-Williams formula used is:

$$P = \frac{4{,}524\, Q^{1.85}}{C^{1.85}\, D^{4.87}}$$

Where:

P = pressure drop or friction loss (psi/ft)

Q = flow (gpm)

C = Hazen-Williams flow coefficient

D = internal pipe diameter (in.)

In actual flow problems, the values of P (in psi/ft or psi/1,000 ft) are precalculated and published in tables for various pipe sizes, flows and C values. Additional tables give values that enable a value of P for a given flow, the C factor and pipe size to be converted to a different flow, C factor or pipe size based on the relationships:

$$P_1 = P_2 \left(\frac{Q_1}{Q_2}\right)^{1.85}$$

$$P_1 = P_2 \left(\frac{C_2}{C_1}\right)^{1.85} \qquad P_1 = P_2 \left(\frac{D_2}{D_1}\right)^{4.87}$$

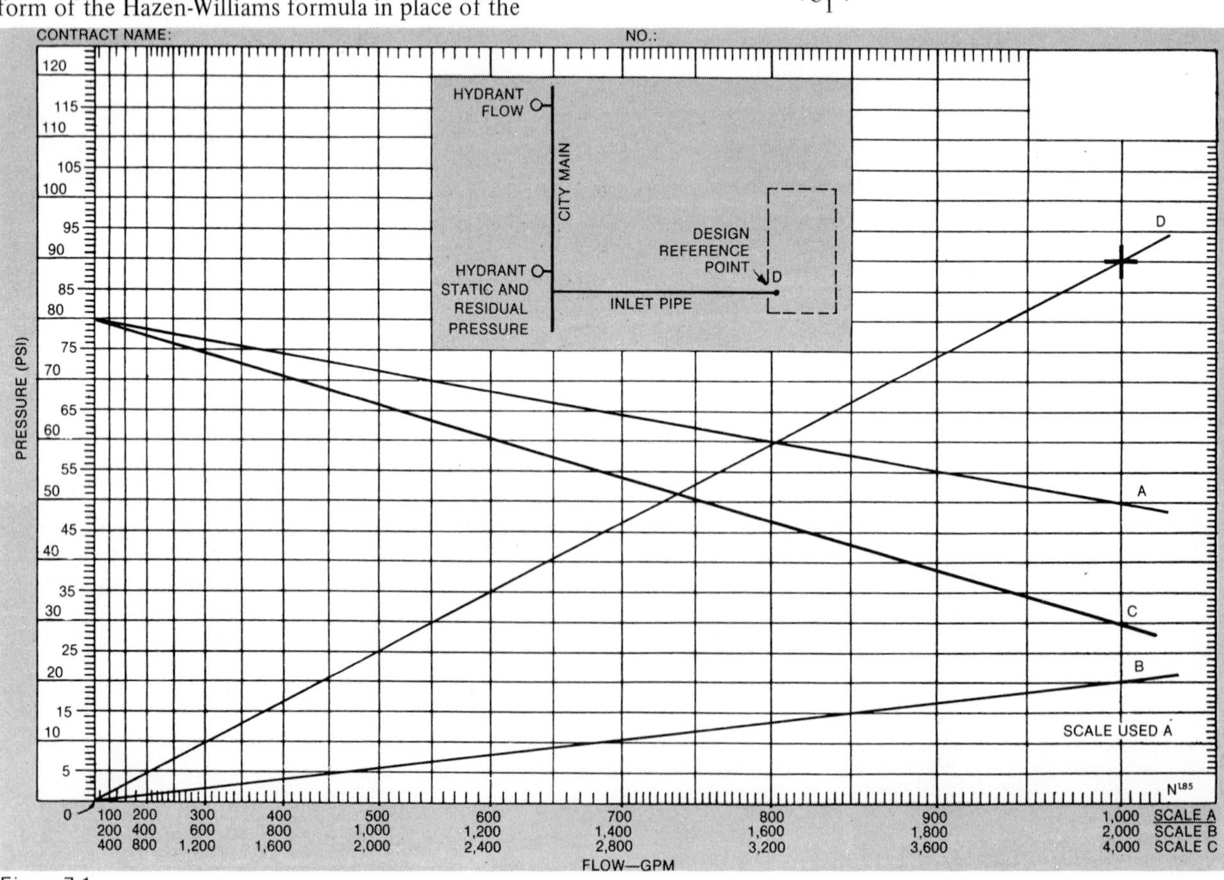

Figure 7.1.

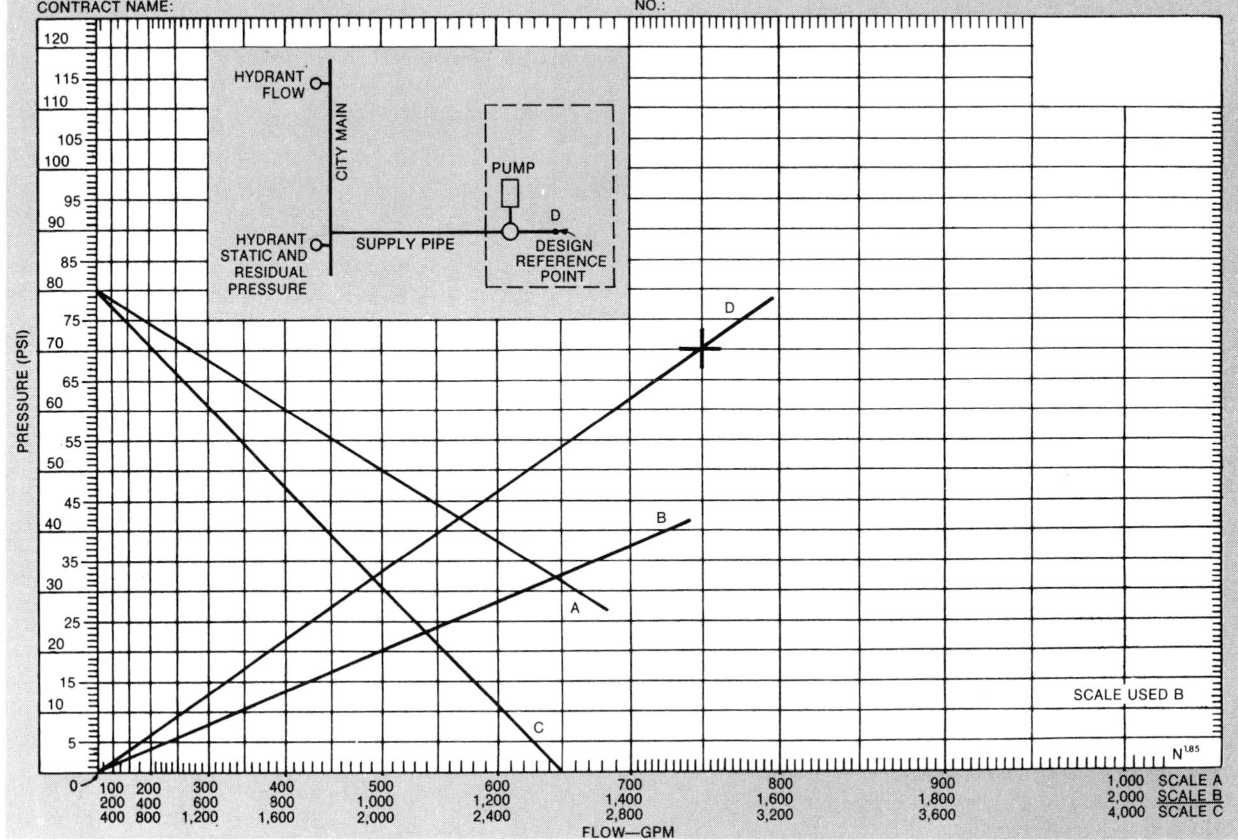

Figure 7.2.

These enable calculations in any system to be simplified by resolving all pipe to one equivalent diameter and C factor.

Simple loops are easily resolved by recognizing that flow divides proportionally to the equivalent lengths of pipe in the loop segments and that pressure drop over each loop segment is the same. Thus:

$$Q_T = Q_1 + Q_2$$

$$Q_T = Q_1 \left[1 + \left(\frac{L_2}{L_1} \right)^{0.54} \right]$$

Where:

Q_T = total flow into loop (gpm)

Q_1 = flow in leg with equivalent length L_1 ft

Q_2 = flow in leg with equivalent length L_2 ft

This method requires that all pipe in both loop legs first be resolved to a single diameter and C factor.

Compound loops and grid systems are analyzed hydraulically by use of Hardy-Cross or some similar system of analysis. Since several computer programs are now available, all but the most ardent yeoman will solve a grid analysis by computer.

In a system where all pipe has been reduced to a single equivalent C factor and pipe diameter, the Hazen-Williams formula can be expressed in a simplified form as

$$\Delta P = KQ^{1.85}$$

Where

ΔP = pressure drop (psi)

Q = flow (gpm)

K = constant combining all other HW factors.

A special semilogarithmic graph paper, which presents ΔP against $Q^{1.85}$, has been developed and is widely used for analysis of fire protection water supply systems. A complete treatment of how to use this paper can be found in *Hydraulic Systems for the Fire Protection Engineer,* by Clyde M. Wood, or in the *Factory Mutual Handbook of Industrial Loss Prevention,* published by McGraw-Hill.

A few of the more basic examples of its value follow.

Figure 7.1 is an analysis of a city supply based on a waterflow test. Curve A is the plot of the actual waterflow test. Curve B is the friction loss from the point of the flow-test pressure values to the design reference point. Curve C is the resultant value of the

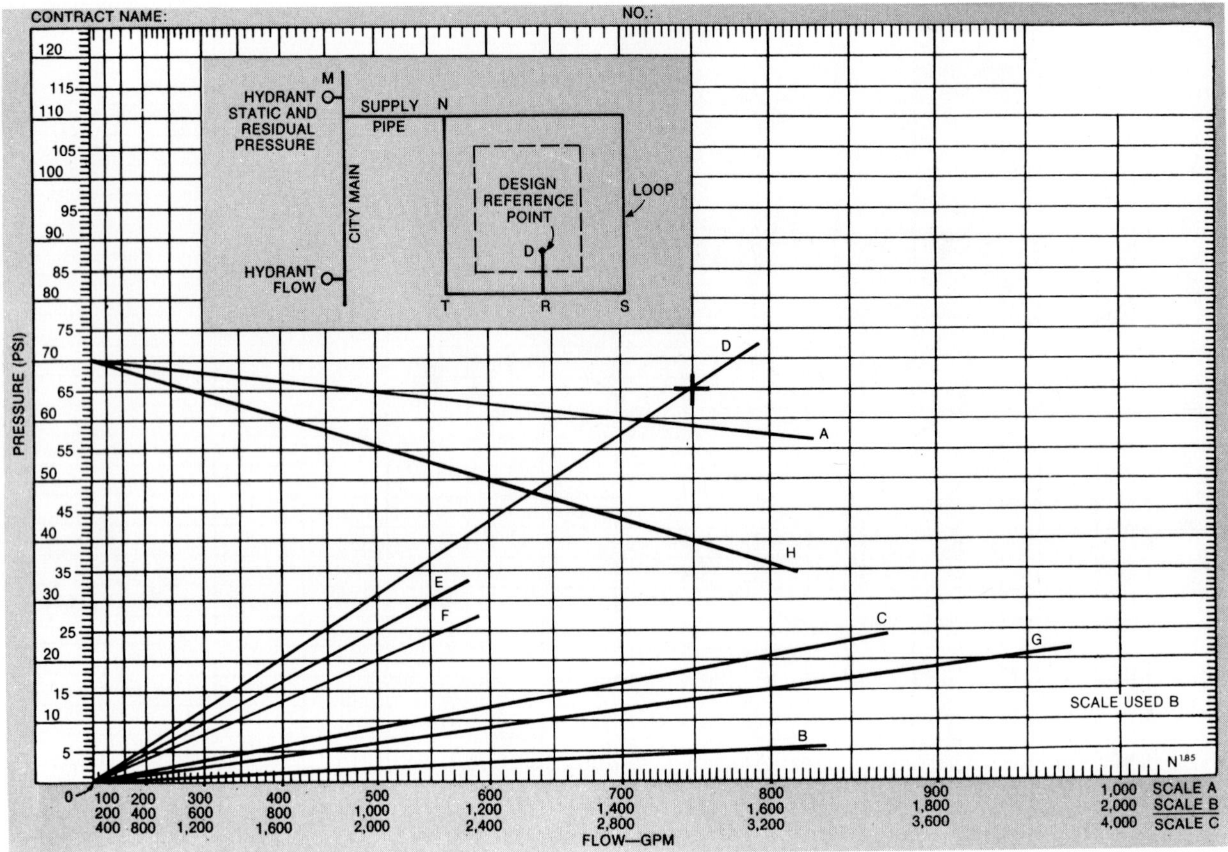

Figure 7.3.

city supply at the design reference point. If Curve D represents the system demand at the design reference point, it is obvious that a pump that provides a pressure boost of 60 psi at 1,000 gpm is needed to develop this city supply.

Figure 7.2 shows this same city supply when a greater demand is required. Curve A is the plot of the actual waterflow test. Curve B is the friction loss from the point of the flow-test pressure values to the design reference point. Curve C is the resultant city supply at the design reference point. If Curve D represents the system demand at the design reference point (i.e., 70 psi at 1,500 gpm), it is obvious that no pump can improve this city system to satisfy the demand. *A pump cannot improve capacity; it can only boost pressure.*

Figure 7.3 illustrates how a simple loop system friction loss and available supply situation can be analyzed. Curve A is the city flow test at point M. Curve B is the combined friction loss of pipe segments MN and RD. Curve E is the friction loss in loop segment NSR and Curve F is the friction loss in loop segment NTR. Curve G is the friction loss over the loop. It can be found by adding the flows from Curves E and F at any selected pressure. Curve C is the total friction loss from M to D over segment MN, loop NR and segment RD. Curve H is the resultant value of the city supply at the design reference point. If Curve D repre-

sents the system demand at the design reference point, it is obvious that a pump that provides a pressure boost of 25 psi at 1,500 gpm is needed to develop this city supply.

Many other examples could be developed to illustrate the use of these graphical methods that often enable a rapid solution to water supply system problems as a tool in designing a new system and in analyzing an existing system.

Specifying and Design Tips

Probably most important in determining system demands is early contact with code, insurance and water department approving authorities. The names of these organizations should be carried forward into the specification. The designer must pin down requirements for piping—pressure, laying conditions, materials, linings, fittings, strapping and thrust blocks—using current NFPA, ANSI, AWWA, federal or other standards. (See NFPA 24 for list.)

Pumps, drivers, controllers and accessory equipment must be specified in a separate section.

The designer should specify reservoirs or tanks, hydrants, valves and other miscellaneous equipment by performance and quality.

Finally, the designer should not forget to list items needed for orderly expansion, when needed.

Fire Pump Design
and Applications

CHARLES B. MILLER, P.E.

Pumps and pumping systems are routinely designed and installed every day by the design engineer. What, then, is so distinctive about pumps, pump controllers and their drivers when used in fire protection service? The designed-systems approach, contrasted with the code-book approach, calls for recognition of certain differences between "fire" pumps and other pumps and equipment.

Industrial, water supply, and other pumps normally operate regularly. Fire pumps, however, have only brief periods of test operation and rare periods of operation in fire extinguishment, yet must be reliably capable of instant starting and running after days, weeks, or even months of inactivity. Maintenance is reasonably assured for pumps in daily use, but, although they deserve it, the fire pumps may not be maintained regularly.

Fire pumps must operate in fire emergencies that occur at the same time as other catastrophes, such as tornadoes, hurricanes, lightning storms, riots, and when protected properties are closed or lightly attended. Other pumps may have design and devices to protect them from short circuits, low voltage or other damaging conditions, but the fire pump must be designed to be kept running under any condition, including sacrificing itself, when providing water for fire protection.

Every pump, driver, controller, or accessory must be specifically approved for fire pump service by Underwriters Laboratories, Inc., or Factory Mutual Engineering Association. These approvals require a distinctive fire-pump curve producing not less than rated total head at rated capacity, and 150 percent of rated capacity at a total head not less than 65 percent of total rated head. Also, the shutoff total head should not exceed 120 percent of total rated head for horizontal pumps and 140 percent for vertical-shaft turbine-type pumps.

The History of Fire Pumps

Early fire pumps were commonly reciprocating types, driven by steam power, in a day when large industrial properties had to maintain a reliable steam supply but electric power was subject to frequent interruption. Improvements in generating facilities, local distribution and protective control systems and interconnected systems have made electric power almost a standard power source for the centrifugal fire pump. Only conditions such as tornadoes and hurricanes, interruptions due to riot, or a radial feeder situation (a spur from a network system without alternate source of supply) generally impair reliability of electric power.

Steam-powered fire pumps, then electric- and gasoline-powered ones, were operated manually for many years. Today most fire pumps are controlled automatically.

Internal combustion engine drivers were long considered as supplemental units only or used where other sources of power were undependable or unavailable. Today, the compression ignition diesel engine is con-

sidered one of the most dependable sources of power
for driving fire pumps.

Increased combustibility of many industrial products,
the economy of large, open manufacturing areas, and
the high-piled storage made possible by modern materials
handling equipment have combined to demand a greater
volume of water at higher pressures for adequate fire
protection. Height and economy limit the ability of
elevated tanks to meet these additional requirements;
public water supplies often are inadequate and require
supplementing. The fire pump, however, is flexible
and nearly ideal when well designed.

Horizontal Pumps

The standard horizontal fire pump, driven most frequently by electric motor or diesel engine, is the unit
most commonly used for fire protection. Experience
has demonstrated the unreliability of pump suction
under lift with the necessary complication of priming
arrangements. The above-ground reservoir provides
suction under head, economy and any reasonable
quantity of water required.

Use of an eccentric reducer avoids entrapment of air
in the level suction pipe and spills it into the upper
casing of the pump where an automatic air release vents
it to the atmosphere, thus avoiding loss of efficiency
in the centrifugal pump impellers.

A relief valve is required when the maximum head
of the pump, plus maximum possible suction head,
exceeds the safe working pressure of the fire protection system (usually 150 psi except in some standpipe
systems). This valve must always be provided for an
internal combustion engine or other adjustable speed
driver.

In most cases, automatically controlled fire pumps
must have a 3/4-inch circulation relief valve set slightly

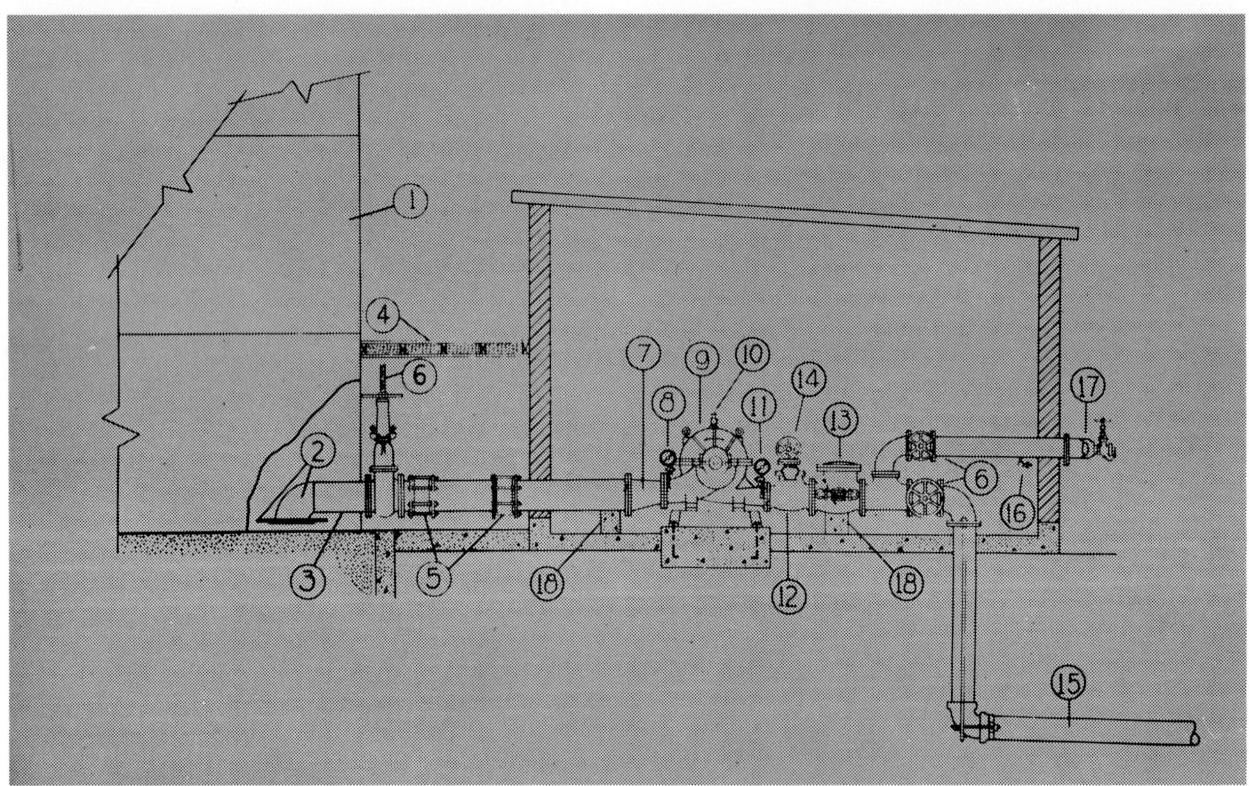

NFPA No. 20, Fig. 100 a. Centrifugal Fire Pump Installation where Pump Takes Water Always Under a Head.

1. Above Ground Suction Tank.
2. Entrance Elbow and 4 x 4 ft., square vortex plate, 4 in. above bottom of tank.
3. Suction Pipe.
4. Frostproof Casing.
5. Flexible Couplings.
6. O.S. & Y. Gate Valve.
7. Eccentric Reducer.
8. Suction Gauge.
9. Horizontal Fire Pump.
10. Umbrella Cock or Automatic Air Release
11. Discharge Gauge.
12. Reducing Tee.
13. Discharge Check Valve.
14. Relief Valve (if required).
15. Discharge Pipe.
16. Drain Valve or Ball Drip.
17. Hose Valve Manifold with Hose Valves.
18. Pipe Supports.

below the pump shutoff pressure to keep cool water flowing through the pump to protect it from overheating when operating with no discharge.

Manifolded hose valves, located outside the pump room where possible, permit measured flow of water

in periodic running tests to prove maintenance of the fire pump's pressure head and volume, certified at the time of the original installation. However, many modern installations are designed with "built-in" orifice plate or venturi meters to simplify the task of routine testing.

For fire protection systems in tall buildings or where water pressures may exceed 125 psi, water hammer may be severe when the automatic controller shuts down the fire pump. In such cases an antiwater-hammer type (sometimes called silent check valve) may be considered for the discharge check valve.

The Booster Pump (Low Pressure 40–100 psi)

The booster fire pump is used where a public water supply provides a sufficient amount of water for the fire protection system but with pressure at this volume too low for fire protection requirements. The volume and head of the pump are designed to boost the available water pressure up to the required head at the required volume.

From the eccentric suction reducer through the pump and discharge connections, the design of the booster pump follows the same principles as the standard fire pump, with two exceptions: First, if the suction supply is under sufficient pressure to be of material value to the fire protection system without the pump, then the pump is installed with a bypass, as in Figure 8.1 (NFPA 20, Figure 143e). If for any reason the pump is inoperative or shut off for repairs, the available water supply can reach the fire protection system through the bypass.

Second, since existence of a material pressure in the suction supply may affect the horsepower requirement of the fire pump electric motor, the pump specification must include the static pressure of the suction supply and its flowing pressure at 150 percent of the pump's rated capacity.

The Jockey Pump

A uniform and relatively high static pressure on most fire protection systems should be maintained to give that pressure instantly for the first water flow in case of fire, to test the integrity of the piping system permanently, and to avoid frequent starts and runs of the fire pump. If another water supply (public main, elevated tank or pressure tank) automatically maintains a sufficiently high static pressure, a jockey (pressure maintenance or loss-leakage) pump is not necessary.

Without another such water supply, frequent starting

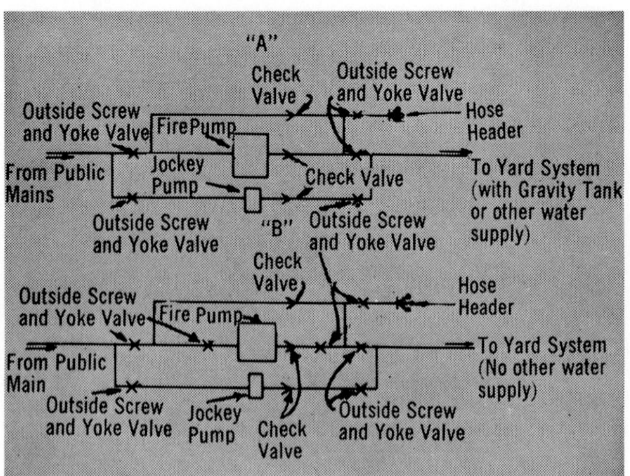

Figure 8.1.

NFPA No. 20, Fig. 143 e. Schematic Diagram of the Suggested Arrangement for a Booster Pump with a Bypass.

Under Condition A, where another water supply stands by to supply the fire protection system during pump repairs, fewer control gate valves are required in this arrangement. Extra gate valves are required under Condition B, with no other water supply, to permit use of the hose header as emergency inlet from a fire department pumper to the fire protection system.

of an automatic fire pump of relatively large capacity to maintain the static pressure (some loss by leakage has to be considered normal in a piping system) is expensive. In addition, such practice may eventually reduce the efficiency of the pumping unit and even damage the fire protection system (e.g., through continued water hammer). A small automatic jockey pump (usually 15 to 50 gpm) will maintain the required static pressure. Large volume systems in high-rise buildings rarely need jockey pumps. (See Figure 8.2., NFPA No. 20, Fig. 100a-1.)

The Vertical-Shaft Turbine Fire Pump

Horizontal-shaft pumps are not recommended where the level of the suction water supply is below the level of the pump. The vertical-shaft turbine-type centrifugal pump is designed for this condition. A column pipe (see Figure 8.3, NFPA 20, Figure 200b) supports and suspends its rotating impellers well below the lowest possible standing-water level and conducts the discharge

flow to above-ground level. No priming arrangement is required, because the pump is submerged in water.

This type of fire pump can be used with a wet pit or a bored well as water supply, though design conditions are quite different. A wet pit collects the screened flow of water from a lake, reservoir or stream and in carefully investigated situations may provide a large, reliable and economical volume of water for fire-pump suction. The second impeller from the bottom of the pump bowl assembly (see Figure 8.3, NFPA 20, Figure 200b) must be submerged below the lowest standing-water level in the open body of water supplying the wet pit. For each 1,000 feet of elevation above sea level this minimum submergence must be increased by one foot.

The pump driver may be a vertical hollow-shaft electric motor (see Figure 8.3, NFPA 20, Figure 200b) or right-angle gear drive with diesel, gasoline or natural gas engine, or steam turbine. The speed of the pump must not exceed 1,800 rpm under any conditions. A motor or gear must be equipped with an antireverse ratchet to prevent reverse rotation of the pump impellers. Automatic starters for engine drivers must have an antidieseling device to prevent reverse operation from self-ignition during compression.

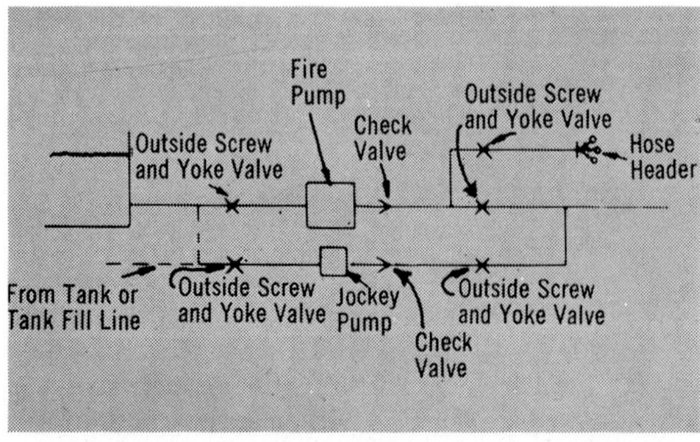

Figure 8.2.

NFPA No. 20, Fig. 100a-1. Jockey Pump Installation with Fire Pump.

NOTE 1: Jockey pump usually required with automatically controlled pumps.
NOTE 2: Jockey pump suction may come from tank filling supply line. This allows high pressure to be maintained on fire protection system even when supply tank may be empty for repairs.

The discharge head and discharge connections are the same as shown in Figure 8.3, NFPA 20, Figure 200b. A relief valve is required for electric motor-driven pumps only when the maximum possible pressure exceeds that for which the fire protection system is designed but is always required for engine- or steam turbine-driven pumps.

The water entrance to the wet pit must be protected against materials that might clog the pump. This protection is obtained with double removable intake screens having an effective net area of openings below lowest standing-water level of one square inch for each gpm at 159 percent of rated pump capacity. A brass or copper-wire screen of 1/2-inch mesh and No. 10 B. & S. gage wire secured to a vertically sliding metal frame is serviceable. It should be arranged so one screen at a time can be removed for cleaning or repairing. In some cases, a rack may be required to protect the screens from damage by heavy floating objects such as logs or timber. In addition, a cone or basket-type strainer (see Figure 8.3, NFPA 20, Figures 200a and 200b) must be attached to the suction manifold of the pump.

The water level in a wet pit supplied from a reservoir is stable, predictable and visible before and during any period of operation of the fire pump. A well bored below the surface of the ground may be the opposite on all three points, but it offers enough possibility of a great volume of water for pump suction to be a challenge to good engineering design and to the experienced well driller.

The static water level in the ground during dry seasons must first be estimated in order to justify proceeding further; the level then must be proved by test after boring. The same applies to the pumping water level at a discharge of 150 percent of the rated pump capacity. The distance between these two levels is the "draw down" (see Figure 8.3, NFPA 20, Figure 200a) and is created by the drain of the pump discharge on the available ground water (well capacity). This distance may vary from negligible to many feet, requiring the services of competent well drillers, with adequate experience in a given area, to judge the difference. It is essential to establish the presence of the required volume of water at the pumping level; first, to assure the submergence of the pump bowl assembly and, second, to establish the distance from pumping water level to ground level. This also establishes the required length of column pipe and the pump pressure needed to lift the water to ground level and supply the pressure required for the fire protection system. The needed pump pressure also determines the size (horsepower) of the driver.

The second impeller from the bottom of the pump bowl assembly should be ten feet below the pumping water level at 150 percent of rated pump capacity. A suitable water level detector must be permanently installed.

Water in the well must be analyzed for corrosiveness, and, when necessary, the pump must be constructed of suitable corrosion-resistant material, such as bronze or red brass. The well screen (for unconsolidated sand and gravel formations only) projecting below the well casing into water-bearing sand should be of No. 304 stainless steel, except where the chloride content of the water exceeding 1,000 parts/million calls for Monel metal. The development and testing of the well must be done with a test pump and the well cleaned of sand at maximum flow before the new fire pump is put in service; this is done to avoid ruining the pump. A nonferrous cone or basket-type strainer must be attached to the suction manifold of the pump.

The discharge head and discharge connections are detailed in Figure 8.3, NFPA 20, Figure 200a. The discharge head is designed to support the driver and pump column and to conduct the flow of water from the column into the discharge connections. Its foundation must be strong enough to carry the weight of the driver and the entire pump and column when full of water.

Electric Motor Fire-Pump Drivers

A first design consideration for the electric motor has to be a completely reliable power supply. The modern public utility network, with its automatic alternate supplies, is most desirable. If current must be taken from a single power station, the power station should be of noncombustible construction and with an apparatus arrangement that minimizes service interruption. The same principles apply all the way to the motor through substation, distribution cables and transformer room. Where single source conditions do not meet this standard, two or more stations or substations free from possibility of the same conditions interrupting them in an emergency should be used.

The diagrams and data in Figure 8.4, NFPA 20, Figure 430 offer illustrations of proper electric supply design. The voltage at the motors should not drop more than 5 percent below the voltage rating of the motors when the pumps are being driven at rated output, pressure, and speed and the power supply lines are carrying their peak loads.

A squirrel-cage motor is generally used for fire pumps. Because of the motor's high starting torque, the capacity of the generating station, the connecting lines and the transformers should be ample and not cause the voltage

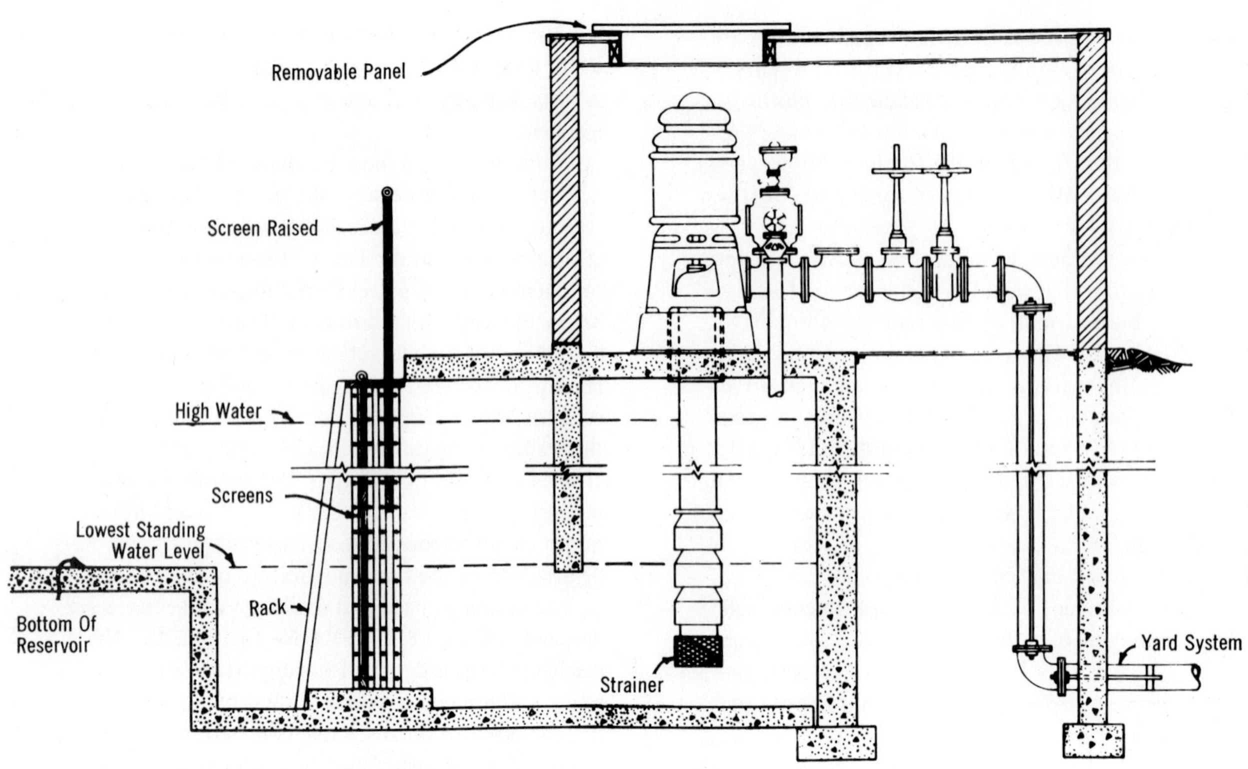

NFPA No. 20, Fig. 200 b. Vertical Shaft Turbine-Type Pump Installation in Wet Pit.

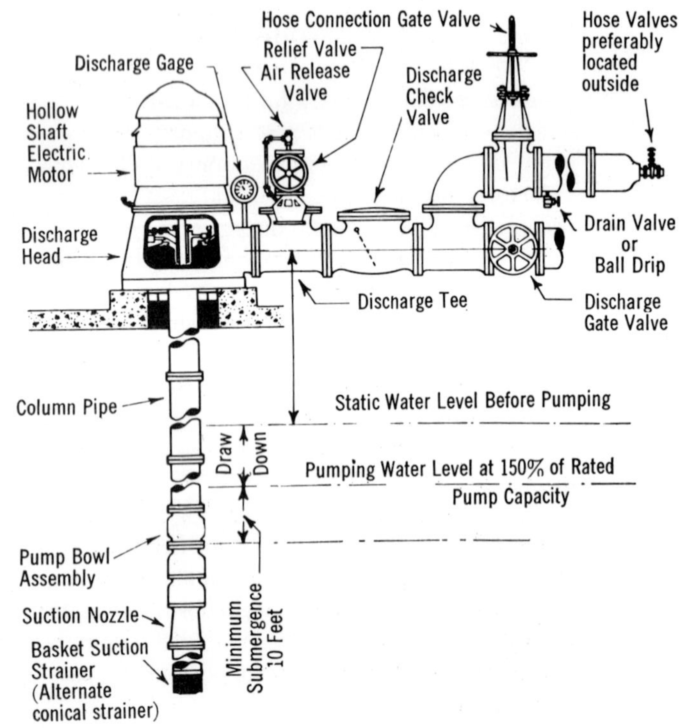

NFPA No. 20, Fig. 200 a. Vertical Shaft Turbine-Type Pump—Installation in a Well.

Figure 8.3.

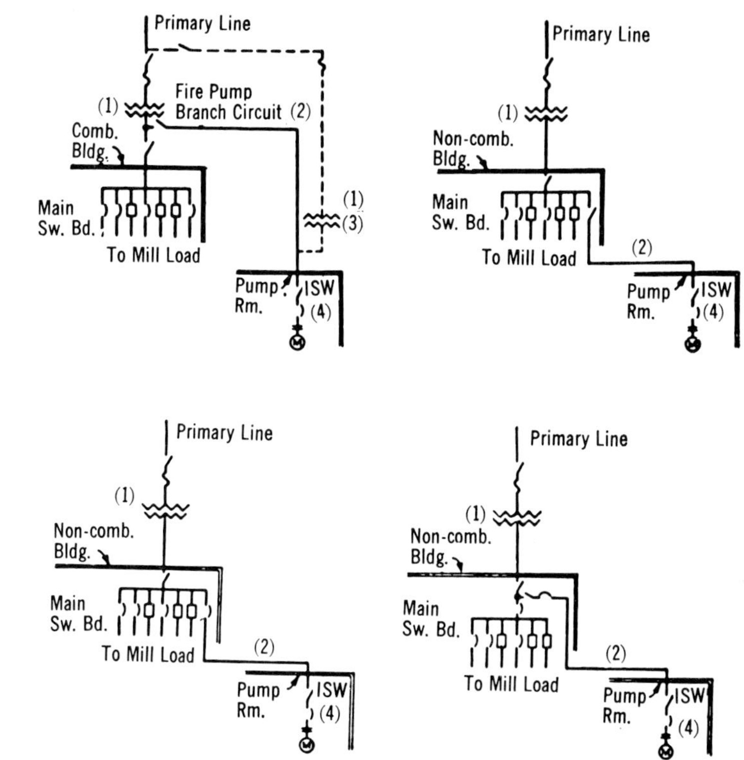

Diagram A—Various Arrangements of Electric Supplies from Public Utility Through Transformers

(1) If primary line is overhead, provide lightning protection for the transformers.

(2) Locate fire pump branch circuit so that it will not be damaged by fire in the protected property. In some cases a circuit breaker may be required by authorities having jurisdiction to protect this circuit and prevent a fault causing the primary fuses to blow. See Article 430.

(3) When main plant transformers are large and would require a circuit breaker of high interrupting capacity in the fire pump controller it may be more economical to provide separate transformers for the fire pump, so that a circuit breaker of lower interrupting capacity may be used in the fire pump controller. For a 75 horsepower motor provide at least 75 kva transformers; for 125 horsepower at least 125 kva.

(4) In some cases where the circuit conditions would permit a circuit breaker of low interrupting capacity, the full load current of the fire pump motor may be high enough to require a circuit breaker which has a higher interrupting capacity.

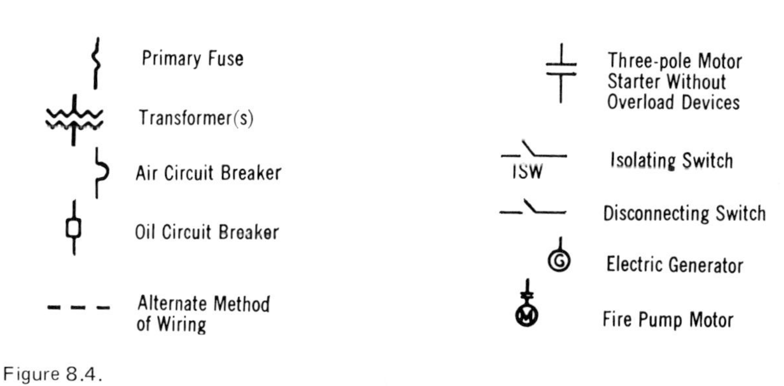

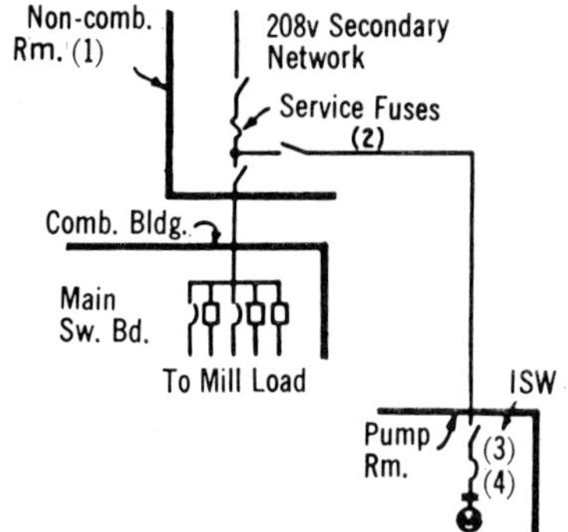

Diagram B.

(1) This room should be accessible in event of fire involving building containing main switchboard so that disconnect switches may be operated to maintain circuit to fire pump motor.

(2) Locate fire pump branch circuit so that it will not be damaged by fire in the protected property. In some cases a circuit breaker may be required by authorities having jurisdiction to protect this circuit and prevent a fault causing the service fuses to blow.

(3) Interrupting capacity of this circuit breaker must be obtained from the utility company.

(4) In some cases where the circuit conditions would permit a circuit breaker of low interrupting capacity, the full load current of the fire pump motor may be high enough to require a circuit breaker which has a higher interrupting capacity.

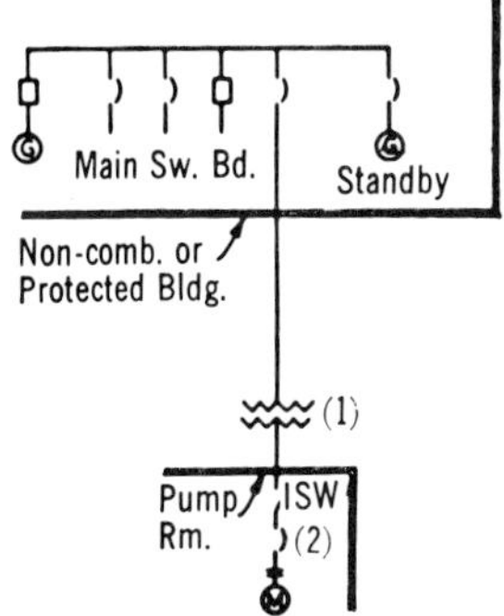

Diagram C.

(1) If fire pump branch circuit is overhead, provide lightning protection. For 75 horsepower motor provide at least 75 kva transformers; for 125 horsepower at least 125 kva.

(2) In some cases where the circuit conditions would permit a circuit breaker of low interrupting capacity, the full load current of the fire pump motor may be high enough to require a circuit breaker which has a higher interrupting capacity.

to drop enough to prevent the motor's starting (not more than 10 percent below normal voltage).

The principle of securing the fire pump from any interruption during an emergency extends to the use of fuses or circuit breakers as protective devices. If installed in the power-supply circuits at utility plants, substations or plant load-distribution centers ahead of the fire-pump feeder circuits, the protective devices must indefinitely hold stalled rotor current conditions of pump motors under maximum pump load. Fuses are not recommended in the fire pump-feeder circuit, and circuit breakers must indefinitely hold stalled rotor current of the motors and other associated fire-pump installation electrical accessories.

The electric motor must be rated for continuous duty and have such a capacity that, at rated voltage (and on AC motors at rated frequency), its full-load ampere rating will not be exceeded (except as allowed by the service factor stamped on the nameplate) under any conditions of pump load. Motors used at altitudes above 3,300 feet must be operated or derated according to NEMA Standard MG1-14.14 (1963). AC motors may be of the squirrel-cage induction-type unless their high-torque starting characteristics are objectionable to the public utility, in which case a wound-rotor-type may be necessary. Squirrel-cage induction motors should have normal starting and breakdown torque; the locked rotor, current of 3-phase, constant-speed induction motors must not exceed values specified for fire-pump service.

Drip-proof motors usually are specified for fire pumps because they are protected against the impact of dripping or splashing water (as from pump leakage) on windings or other energized mechanisms. Open motors subject to possible water splash from hose connections close to the pump must have a splash partition between the pump and motor. Splash-proof or totally enclosed, fan-cooled motors are suitable for fire-pump service where conditions may require them, but the relation of their nameplate service factor to allowable motor overload has to be considered; sometimes a larger motor is required.

**Internal Combustion Engine
Fire-Pump Drivers***

The most reliable kind of control, ignition, fuel (including fuel supply), starting operation and running operation are important considerations in selecting the type of internal combustion engine driver for a fire pump. These factors combine to make the compression-ignition diesel engine a dependable source of power. Spark-

Note: Natural gas and gasoline engines were deleted from NFPA 20 in the 1975 National Fire Codes.

ignition engines may qualify as supplementary sources of power, but first fuel preference should be given natural gas, followed by gasoline.

The engine should have a bare engine brake horsepower rating at least 20 percent greater than the maximum brake horsepower required to drive the fire pump at rated rpm. This hp rating allows tolerance in rating new production engines, operation of accessories, and reserve for reliability of performance and depreciation with age and use. For each 1,000 feet of elevation above sea level, the engineer should allow a deduction of 5 percent of the power shown on the curve of the engine. (Note: Engines rated by UL and FM are derated in testing by the approval agency.)

The engine should have a governor to regulate its speed within a range of 10 percent between shutoff and maximum-load condition of the pump. The governor should be set to maintain rated pump speed at rated pump load. The tachometer, oil pressure gauge and temperature gauge should be mounted on a control panel at the engine. Main battery contactors should be manually operable.

Compression-ignition diesel and spark-ignition gasoline or natural-gas engines are started by an electric device taking current from a storage battery. This highly essential function of starting, especially automatic pump units (which always have two batteries), emphasizes the need for a good quality, ample-capacity battery and equally good quality automatic recharging equipment to maintain the battery in top condition.

The capacity and arrangement of diesel or gasoline fuel supplies also follow emergency-equipment design principles. Among the considerations are minimum storage capacity to operate each engine at least eight hours (one gal/hp is a good rule of thumb), a separate fuel tank and fuel line for each engine, interconnection and valving of multiple fuel lines so that all engines are operable even though one or more fuel tanks may be out of service, a fuel installation in compliance with safety standards and ordinances, and a fuel piping arrangement according to the engine manufacturer's specifications.

When used, every feature of a natural-gas fuel supply should be analyzed for reliability. A prime requirement is that restrictions applied by some utility companies to other uses of gas cannot stop the supply to the fire pump. Supply piping must be adequately sized to maintain required pressure at the engine during maximum demand for other uses, whether in the utility mains or within the boundaries of the private property. A common allowance is twelve cubic feet of 1,000 Btu natural gas/hp/hr. To produce rated hp, the engine must be supplied with gas having a Btu value equal to or greater than the Btu value specified by the engine manufacturer. Where lower gas Btu values are encountered engine hp must be derated accordingly. This may require one to specify a larger engine to meet the power requirement of the pump.

Figure 8.5 (NFPA 20, Figure 644) shows a typical natural-gas fuel system supplying both fire pump and plant service. This system has an automatic low-pressure cutoff valve to preserve an uninterrupted supply to the fire pump in case of a break in the plant service line. The pressure regulator at the pump reduces the available gas pressure to the optimum low pressure for the engine carburetor. The safety solenoid valve is electrically opened when the engine ignition is turned on and it closes automatically when the ignition is turned off. In case of malfunction of this valve, there is a bypass around it that has a valve for manual opening in emergency. (This emergency valve should have a visual or audible signal to show when it is open.) The electrical system of the engine powers all electric controls. Gas supply piping outside the pump house should be installed with pitch to drain to avoid water traps or pockets that might block gas flow. The outside shutoff is for manual operating and should be locked open with a breakable lock.

Fire-Pump Controllers for Electric Drive

Inclusion of controllers in the Fire Protection Equipment List of Underwriters Laboratories, Inc., or in the Factory Mutual Approval Guide certifies numerous detailed features essential for fire-pump service. The design engineer should pay careful attention to considerations other than the previously mentioned squirrel-cage induction motors and their high-torque starting.

Across-the-line starting is preferred in a fire-pump controller, but if the combination is not acceptable to the power company, the engineer can substitute a primary resistance reduced-voltage type of controller to reduce the starting shock to the power system. To assure the controller's ability to operate under extreme conditions, its circuit breaker must have an interrupting capacity equal to or greater than the maximum possible short-circuit current at location of pump in the particular circuit in which it is used (in no case less than 14,000 amp symmetrical).

Where two or more motor-driven pumps are automatically controlled, there must be a sequential timing device in the controllers to start the units at intervals (usually of not less than five seconds duration) so each will reach full speed before the next starts. This timing avoids an excessive starting load.

Controllers may be designed to stop fire pumps automatically (by a water-pressure control switch) or manually. Manual stoppage may be preferred to keep the pumps running with full pressure available throughout any fire emergency. When the controller is arranged for automatic shutdown, automatic response to system pressure may cause starts and stops to be close enough together that the motor and starting resistors are overheated. The controller, therefore, has a running-period timer to override this condition. The timer is set to keep the motor running not less than one minute for each ten hp of motor rating, not to exceed seven minutes.

Each controller's automatic water-pressure control switch is piped to the fire pump discharge as shown in Figure 8.6, NFPA 20, Figure 515d-1. The arrangement of globe valve and pet-cock drain between two restricted orifices makes it easy to test the automatic start-and-stop operation of the controller and the pump.

Controllers may use contacts for any or all of the following options:

1. *Remote alarm* to signal that the pump is in operation

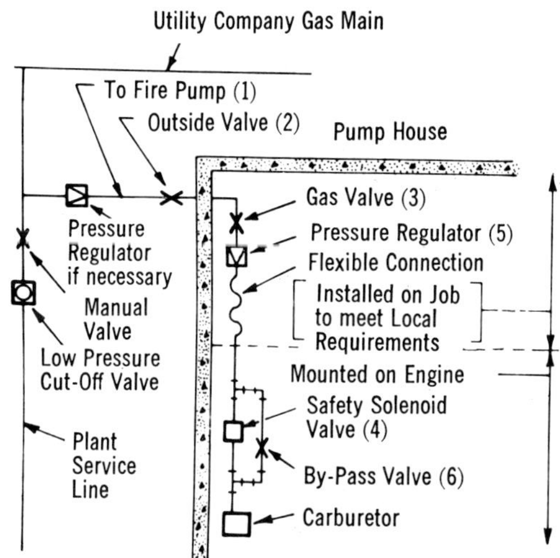

NFPA No. 20, Fig. 644. Typical Natural Gas Fuel System

1. Outside gas pipe to be iron or steel; indoor pipe to be steel, wrought iron, copper or brass.
2. Outside valve needed on pipe over 2½ in. diameter.
3. Approved gas valve locked open or supervised.
4. Approved, electric-opening, self-closing valve.
5. Install as close to engine as possible.
6. Valve to be provided with visual or audible signal to show when it is open.

Figure 8.5.

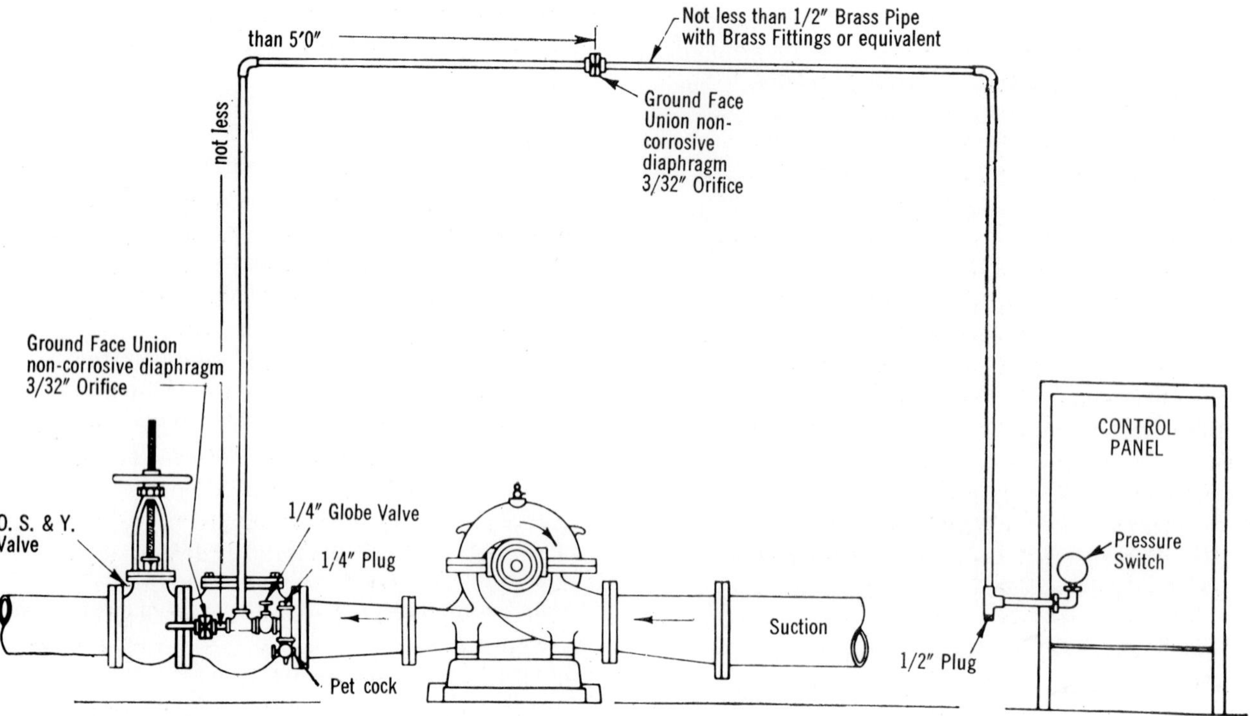

NFPA No. 20, Fig. 515d-1. Piping Connection for Automatic Pressure Switch.

Figure 8.6.

or that there is trouble on the controller and/or pump.

2. *Audible- or visual-alarm* energized by another reliable source of power supply, through a drop-out relay on the controller, to signal loss of power for pump operation.

3. *Remote manual push-button* arranged to start, but not stop, the pump.

4. *Instantaneous starting of the pump* from operation of automatic-sprinkler deluge alarm or dry-pipe valves by closed-circuit connection from the valves to the controller through a drop-out relay.

Fire-Pump Controllers for Internal Combustion Engine

Experience in starting, driving, and maintaining an automobile gives some indication of the comprehensive task of a piece of equipment responsible for automatically performing those functions in a fire pump engine. Some models are designed for the diesel engine starting, others for the somewhat different gasoline starting.

Either the water-pressure control switch, a power-failure relay (upon loss of AC power to battery chargers and timers), remote manual control, or weekly program timer (if used) will engage the controller's engine cranking and running circuits. When the engine is running under its own power, the cranking circuits are disconnected and the running circuits are maintained. If the engine does not start after approximately thirty seconds, cranking stops briefly, the controller switches to a second battery and continues to crank. If this on-off and battery switchover has continued for about ninety seconds without the engine starting, cranking is discontinued, an engine-failure alarm sounds, a signal light and remote-failure signal are energized. If one of the two batteries is nonoperative, the control locks-in on the other battery during the cranking sequence. While the engine is running, either low oil pressure in the lubrication system or high engine-jacket water temperature will sound the engine-failure alarm. When required on spark-ignited engines, the controller has circuits to operate chokes or similar devices.

If the pump room is not constantly attended, all the aforementioned trouble signals, as well as a separate pump-running signal and separate indicator to signal when the controller main switch is in "off" or "manual" position, can be located remotely.

Engine controllers are normally arranged for automatic start and manual stop, though automatic stop is possible for an unusual condition.

When reliable maintenance service is available, the fire pump should be started once a week, run until warm, and manually stopped. This test operation is necessary to prove the good condition of pump, engine, fuel supplies, batteries, and other starting equipment. If such service is not available, the controller should have a weekly program timer and usually a recording pressure gauge to perform the testing automatically and provide a record of it.

Controllers for internal combustion engines should have built-in supplementary automatic battery charging equipment capable of maintaining starting batteries in a fully charged (not overcharged) condition. Controllers for multiple pump units should have a sequential timing device to prevent any one pump starting simultaneously with any other. Controllers for spark-ignited engines should have circuits to operate anti-dieseling devices unless it is established that they are not needed for the engine.

If engine-driven pumps are associated with ones driven by electric motors, their controllers may be arranged to start automatically in the event of a power interruption to the motor-driven pumps.

Only salient features of fire-pump design have been presented in this chapter. For more specific details of fire-pump systems, mechanical/electrical engineers should consult publications listed in the references.

References

NFPA No. 20 *Centrifugal Fire Pumps,* published annually by the National Fire Protection Assn., Boston.

Handbook of Industrial Loss Prevention (2nd edition, Chapter 20, "Fire Pumps"), Factory Mutual Engineering Corp., McGraw-Hill Book Co.

Fire Protection Equipment List (section titled "Pumping Equipment for Fire Service"), published annually by Underwriters Laboratories, Inc., Chicago.

Factory Mutual Approval Guide (section titled "Fire Pumps and Tanks"), published annually by Factory Mutual Engineering Corp., Norwood, Mass.

Standard of Hydraulic Institute, published by Hydraulic Institute, New York City.

American Standard Specifications for Deep Well Vertical Turbine Pumps, published by American Water Works Assn., Inc., New York City.

Wells, Department of the Army and the Air Force, Superintendent of Documents, U.S. Government Printing Office, Washington, D.C.

Sprinkler Systems

WAYNE E. AULT, P.E.

The proper design of automatic sprinkler systems can favorably influence the cost of building construction, make hazardous processes feasible and reduce the risk of loss of valuable goods. Modern specification techniques accommodate automatic sprinkler systems early in the design phase, resulting in a less expensive fire protection package. Water supply, electrical power supply, and signaling facilities can be integrated with other facilities, thus preserving space and funds.

When "fire prevention" fails and "fire extinguishers" are depleted, the operation of the automatic sprinkler system rapidly and efficiently corrects for unsafe deviations in building construction, occupancy operation, or limited fire department response. Continuity of operation, profits, employment, and fire protection improve when complete automatic sprinkler systems are installed. Insurance costs less. In one case, an insurance company lowered its premium by 85 percent when a sprinkler system was installed in one of its client's plants. The reduced premiums paid for the sprinkler system installation in less than four years and subsequently availed approximately $20,000 each year for other uses [1].

When human safety is of prime consideration, the early detection, alarm, and extinguishment of fires provided by automatic sprinkler systems warrants serious consideration. Rarely are lives lost in sprinklered buildings. In the few reported cases, unusual circumstances were involved, such as when individuals were trapped in early or rapidly developing fires. One report of an incident attests to the security provided by automatic sprinkler protection: "Evidence indicated that had it not been for the prompt operation of the automatic sprinkler system, the six girls probably would have perished. The man who died had been too quickly enveloped in fire [match factory] to be saved by the sprinklers."[2]

The misconception that sprinkler discharge can overcome or "drown" occupants of a room has been dispelled by actual demonstration on many occasions. The water discharge rate equals one-fiftieth that of a shower head [3], and the amount of steam generated in putting out a fire is less than that generated by hose streams.

How They Work

Automatic sprinklers open at a predetermined temperature. The heat-sensitive operating elements may be fusible metal alloys, organic softening materials or expandable organic liquids in frangible glass ampules. Speed of operation is dependent on the mass of the operating elements. Sufficient potential releasing energy is built in to overcome the effects of long-time exposure to most industrial atmospheres. While it is usually desirable that operation be as prompt as possible, normal room temperatures must be considered. The following table [4] shows recommended operating temperatures for various ceiling temperatures:

Maximum Ceiling Temperature °F	Temperature Rating °F
100	135 to 170
150	175 to 225
225	250 to 300
300	325 to 375
375	400 to 475
475	500 to 575

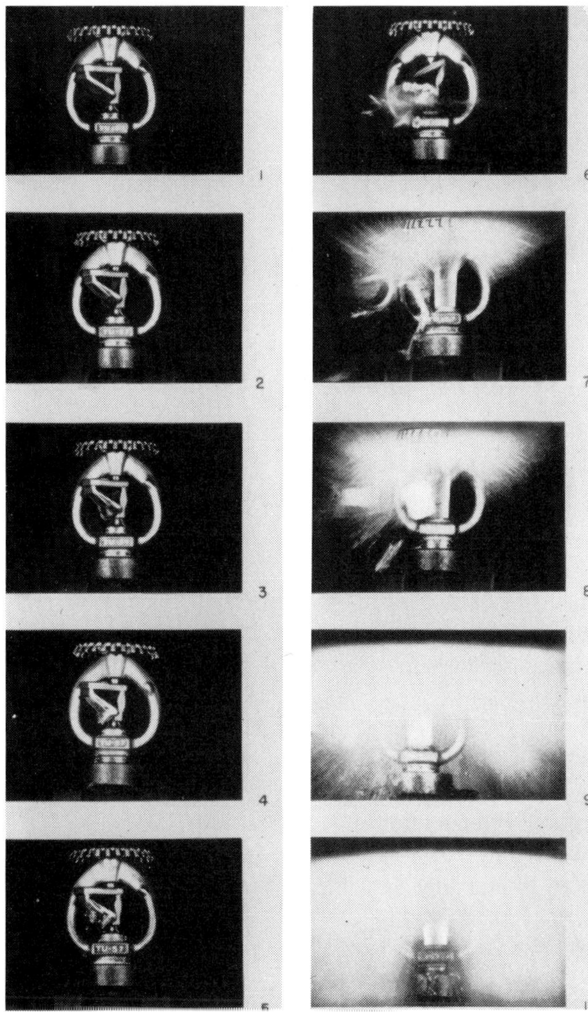

These photos show the operation of a typical solder-type automatic sprinkler. Heat melts the solder; members of the solder link begin to separate (the sloping side of the triangle in photos 1 to 5); the link and lever arrangement completely separate (photo 6); the cap over the sprinkler orifice releases and water escapes and strikes the deflector (photos 7 to 10). (Photos courtesy of the National Fire Protection Assn.)

The waterway of a sprinkler needs a large orifice (usually half an inch in diameter) to produce an effective discharge at low pressures and to pass solids carried by non-potable fire protection water supplies. Other orifice sizes are available for special situations. The 17/32-inch orifice produces 140 percent more water than the half inch. This larger size is particularly useful where large flow rates are required. Small-orifice sprinklers are often used in light-hazard occupancies. Table 9.1 shows nominal discharge capacities of orifice sprinklers [5].

Sprinklers are designed to produce nearly uniform distribution of water over a given area. The solid cylin-

drical flow of water through the orifice strikes the deflector and is diverted and broken into an infinite number of water droplets. The size and velocity of the droplets vary to produce "reach" and uniform wetting of the volume between the ceiling and the floor. Some droplets absorb heat moving through the room; some penetrate fire drafts and reach burning surfaces; others sprinkle adjacent flammable areas to retard spread of fire. Systems are designed with overlapping patterns to provide more efficent fire control when more than one sprinkler operates. Spacing sprinklers too closely, however, is not only wasteful but can also retard prompt operation of needed adjacent sprinkler operation through wetting or "cold-soldering."

Upright sprinklers are most often used in sprinkler systems. Unfortunately, the sight of piping hanging on long rods or of the oval-shaped automatic sprinkler near the ceiling is not aesthetically satisfying. Piping and hanging equipment can be concealed above false ceilings. Pendent sprinklers mounted below the ceiling, provide the same circular coverage as upright sprinklers and are modularly spaced from each other.

Sidewall sprinklers are located adjacent to the wall. Their deflectors are designed to obtain long throw of water toward the opposite wall and additional sprinklers need not be installed to obtain complete coverage of the smaller rooms. This style of sprinkler is particularly useful for the protection of hotel rooms, apartments, and other light hazard occupancies where a minimum of ceiling obstruction is desired.

Aesthetically, the appearance of the sprinkler system can be made pleasing by the use of sprinklers especially designed to reduce intrusion below the ceiling. The so-called low-profile ceiling, recessed, or concealed types are often selected at additional cost.

Performance Records

The life expectancy of an automatic sprinkler is fifty to one hundred years, but once they have been operated, they must be replaced, since the heat-sensitive elements must be assembled in the factory. Periodic testing and regular inspection of sprinklers are necessary to ensure that the efficiency of each sprinkler is unimpaired. They must not be painted. If they are exposed to excessive dust, dirt, or atmospheric corrosion, they may need to be replaced periodically. Special protective coatings can be factory-applied to retard the effects of moderate corrosive atmospheres. The replacement cost of individual sprinklers, however, is low.

A fire in a building protected by an automatic sprinkler system does not necessarily operate every

Table 9.1

Nominal Orifice (In.)	"K"[1] Factor	Percent of Nominal 1/2-inch Discharge	Identification[2]
1/4	1.3–1.5	25	1/2 in. IPT—Pintle
5/16	1.8–2.0	33.3	1/2 in. IPT—Pintle
3/8	2.6–2.9	50	1/2 in. IPT—Pintle
7/16	4.0–4.4	75	1/2 in. IPT—Pintle
1/2	5.3–5.8	100	1/2 in. IPT
17/32	7.4–8.2	140	3/4 in. IPT or 1/2 in. IPT—Pintle

[1] "K" factor is the constant in the formula.

$$Q = K\sqrt{P} \qquad \text{Where } Q = \text{Flow in GPM}$$
$$P = \text{Pressure in PSI}$$

[2] With the exception of 1/2-inch orifice and 17/32-inch orifice, 3/4-in. IPT (iron pipe thread) sprinklers, the nominal orifice size is cast or stamped on the wrench boss of the sprinkler frame.

sprinkler. Fifteen or fewer sprinklers operate in 90 percent of the fires, and a single sprinkler, in many instances, will detect and extinguish the fire. (See Table 9.2 [6].)

Records of automatic sprinkler system performance have been kept by the National Fire Protection Assn. since 1897 (see Table 9.3 [7]). No significant changes in the performance of automatic sprinkler systems have occurred, although the types and severity of hazards have increased. Research and the improvement of standards for fire protection systems have produced systems capable of meeting the new advances in technology.

System Design Concepts

Most essential to automatic sprinkler systems is the concept of 100 percent area protection. Since water supplies and piping limit flow, only some of the sprinklers in a system can operate simultaneously. A sprinkler system must extinguish a fire at its point of origin, or else the rapid spread of the fire will overcome the system by opening too many sprinklers. Fires sometimes originate in areas where causes for ignition or combustibles are least expected to exist. Care must be taken to locate sprinklers under tables and on both sides of partitions, in hallways, and even in places not

Table 9.2. Number of Sprinklers Operating (Cumulative), 1925–1969

Number of Automatic Sprinklers Operating	Wet Systems Per Cent	Dry Systems Per Cent	Unknown Systems Per Cent	Total No. of Fires	Per Cent of Total No. of Fires
1	42.6	20.1	33.1	29,733	37.4
2 or fewer	61.0	32.7	50.0	43,396	54.6
3 or fewer	70.2	41.5	59.8	50,769	63.8
4 or fewer	76.2	48.7	66.7	55,795	70.1
5 or fewer	80.2	53.7	70.9	59,156	73.4
6 or fewer	83.2	57.8	75.0	61,814	77.7
7 or fewer	85.2	61.3	77.7	63,724	80.1
8 or fewer	87.0	64.2	80.3	65,348	82.2
9 or fewer	88.3	66.4	82.0	66,571	83.7
10 or fewer	89.4	68.5	83.5	67,629	85.0
11 or fewer	90.4	70.3	84.5	68,533	86.2
12 or fewer	91.2	72.4	86.1	69,464	87.3
13 or fewer	91.7	73.8	87.0	69,990	88.0
14 or fewer	92.6	75.3	88.0	70,788	89.0
15 or fewer	93.1	76.2	89.9	71,313	89.7
20 or fewer	95.0	81.0	91.5	73,347	92.2
25 or fewer	96.0	84.3	92.9	74,464	93.6
30 or fewer	96.9	86.7	94.2	75,411	94.8
35 or fewer	97.3	88.6	95.0	75,976	95.5
40 or fewer	97.7	90.0	95.8	76,472	96.2
50 or fewer	98.1	91.9	96.7	77,079	96.9
75 or fewer	98.9	94.7	98.0	77,995	98.1
100 or fewer	99.4	96.3	98.5	78,533	98.7
200 or fewer	99.8	99.7	99.9	79,384	99.8
All fires	100.0	100.0	100.0	79,544	100.0
No data or no water				1,881	
Total Number of Fires	54,158	13,217	12,169	81,425	

Table 9.3. Summary of Sprinkler Performance, 1897–1969

	Fires 1897–1924		Fires 1925–1969*	
	Number	Per Cent	Number	Per Cent
Satisfactory	31,388	95.8	78,291	96.2
Unsatisfactory**	1,390	4.2	3,134	3.8
Total	32,778	100.0	81,425	100.0

*For the five-year period 1965–1969 the ratio of satisfactory performance was 95.7 percent.

**This category reflects not only poor sprinkler performance but also incidences when water supplies were shut off or inadequate at the time of fire (36 percent of all unsatisfactory performance.).

normally accessible to people. When buildings are altered additional sprinklers may be needed to accommodate changes.

The probable number of sprinklers that may operate in many types of fire situations has been established by performance records in fires and by experimental work on full-scale fire tests. While fire in an ordinary hazard situation may operate fifteen or fewer sprinklers in 90 percent of the recorded cases, fire in an extra hazard situation (e.g., high-piled cardboard cartons) may operate as many as 25 percent of the sprinklers in a single fire area. Spacing between sprinklers, pipe sizing, size of orifices and water supplies are based on the hazard expected. High flow rate systems have piping sized by "hydraulic calculations." Any change in hazard necessitates a change in the automatic sprinkler system's design.

While the sprinkler industry has standardized the water distribution pattern to be produced, some of today's hazards require special consideration. Test fires in various materials storage and handling techniques have produced the common term, "density," defined as sprinkler flow in gallons per minute per square foot. The usual hazard can be protected with densities as low as 0.1 gpm/square feet; requirements for protection of highly piled material or for highly flammable materials may exceed 0.5 gpm/sq ft. (See Chapter 7 for hydraulic design requirements.)

Location and Layout

The locations of sprinklers are "spotted" in building ceiling plans in accordance with spacing rules [8]. These spacing rules are based on several important considerations:

1. Best location to detect fire most rapidly.
2. Best location to discharge water over the greatest area and accommodate ceiling construction, partitions, and floor obstructions.

3. Required flow rate and available water supply.
4. Class of occupancy (light, ordinary or extra-hazard).

For ordinary hazards, sprinklers may be spaced as far as fifteen feet apart and cover up to 130 square feet per sprinkler. However, construction features, high hazards of occupancy, or characteristics of water supply may require sprinklers to be spaced less than ten feet apart, or cover less than ninety square feet per sprinkler.

Water Supply and Piping

Water is the most important ingredient of the sprinkler system; caution, therefore, must be exercised in specifying the water supply requirement. To design a system, the engineer must evaluate the worst probable fire situation and this situation's demand on the water supply, in addition to the demand by expected fire department operations. The building's height and the hydraulic friction losses in piping connections dictate the minimum required flowing (residual) pressure. Design often calls for water supply piping sizes of six inches or larger and potential water demand of 500 to 5,000 gpm for several hours.

Sprinkler system pipe sizes may be "schedule sized" to reduce hydraulic friction loss to a practical value (see Table 9.4 [9]). The schedules are based on the number of sprinklers supplied, the piping and the hazard of the occupancy. Telescope sizing procedures may dictate starting a system with a six-inch pipe and completing the run to the end sprinkler with one-inch pipe.

Piping may be sized by "hydraulic calculations," particularly where there are hazards or occupancies that will demand larger rates of water flow. This design

Table 9.4.

SCHEDULE FOR LIGHT-HAZARD OCCUPANCIES

1 in. pipe	2 sprinklers	2 1/2 in. pipe	30 sprinklers
1 1/4 in. pipe	3 sprinklers	3 in. pipe	60 sprinklers
1 1/2 in. pipe	5 sprinklers	3 1/2 in. pipe	100 sprinklers
2 in. pipe	10 sprinklers	4 in. pipe	Area Limit

SCHEDULE FOR ORDINARY-HAZARD OCCUPANCIES

1 in. pipe	2 sprinklers	3 in. pipe	40 sprinklers
1 1/4 in. pipe	3 sprinklers	3 1/2 in. pipe	65 sprinklers
1 1/2 in. pipe	5 sprinklers	4 in. pipe	100 sprinklers
2 in. pipe	10 sprinklers	5 in. pipe	160 sprinklers
2 1/2 in. pipe	20 sprinklers	6 in. pipe	275 sprinklers
		8 in. pipe	Area Limit

SCHEDULE FOR EXTRA-HAZARD OCCUPANCIES

1 in. pipe	1 sprinkler	3 in. pipe	27 sprinklers
1 1/4 in. pipe	2 sprinklers	3 1/2 in. pipe	40 sprinklers
1 1/2 in. pipe	5 sprinklers	4 in. pipe	55 sprinklers
2 in. pipe	8 sprinklers	5 in. pipe	90 sprinklers
2 1/2 in. pipe	15 sprinklers	6 in. pipe	150 sprinklers
		8 in. pipe	Area Limit

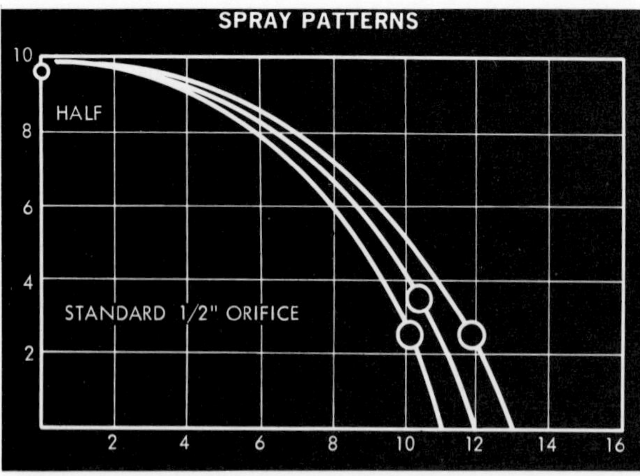

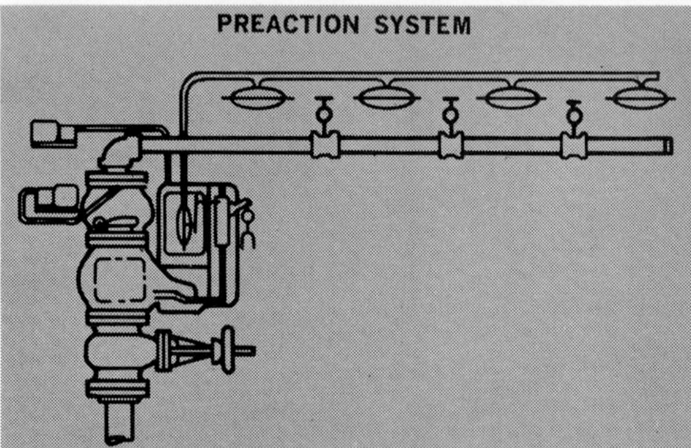

In the preaction system, all sprinklers are closed, and normally there is no water in the piping. At the time of a fire, a supplementary detection system senses the fire and automatically opens a water control valve. This allows water to flow into the piping system. Subsequent water discharge occurs from individual sprinklers as they respond to the heat of the fire. The system shown uses H.A.D. (Heat Actuate Device) detectors and a drop-weight release to open the valve.

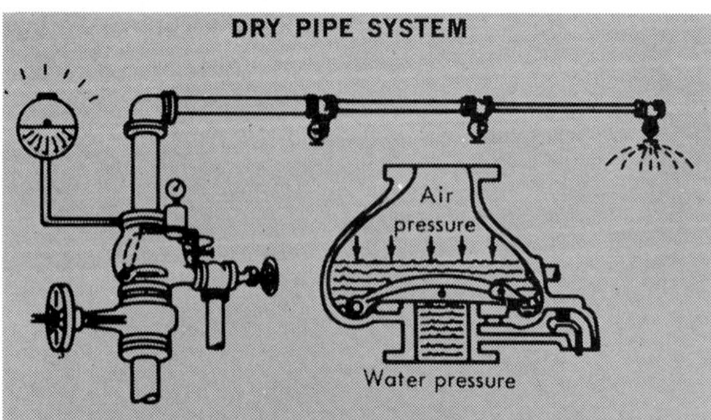

This system uses automatic sprinklers that are attached to a piping system containing air under pressure. When the air is released from the opening of the sprinklers, the water pressure is allowed to open a valve known as a "dry-pipe valve." The water then flows into the piping system and out the opened sprinklers. This system provides protection in areas subject to below-freezing temperatures.

technique assures more uniform distribution of water to sprinklers and conserves pressure and volume. The object of hydraulic calculations may be to reduce the cost of the piping system through reduction in piping sizes by "looping" or "gridding."

Hydraulic design criteria are developed in the National Fire Protection Association Standards for Installation of Sprinkler System, No. 13, and in various other NFPA Occupancy Standards like those for Rack Storage of Materials, No. 231C, and Indoor General Storage, No. 231. The requirement for water densities (gallons per minute per square feet), minimum number of sprinklers expected to operate, simultaneously, over the probable area of fire involvement are based on fire test and experience. A series of calculations, utilizing computers, friction loss tables, special slide rules, or calculators establishes the optimum sizes of pipe gridwork; total water demand and pressure requirements can be established for the water supply. It may be necessary to augment a city water supply with a pump and water storage tank. (See Chapter 7 for designs of water supplies.)

Four Basic System Types

The wet-pipe system is the simplest automatic sprinkler system. Water is constantly present in each automatic sprinkler, and when the sprinkler opens, the fire is immediately sprayed. Only those sprinklers nearest the fire will go into operation. The wet-pipe system should be used where water damage will be negligible and operation time will contain a fire of ordinary combustion. Because the piping is constantly filled with water, this system should not be used where freezing conditions are likely to occur.

The dry-pipe sprinkler system is used in unheated buildings. The piping is filled with air under pressure. When fire occurs, the sprinklers open, relieving the air pressure and admitting water into the piping. Because of the delay between opening of sprinklers and the arrival of water, additional sprinklers are expected to operate.

The air supply must be of a size capable of rapid resetting of the dry-pipe sprinkler system—an automatically controlled electrical air compressor plant air supply. The compressor capacity in cubic feet per minute should be large enough to rapidly pump up the system so protection can be restored in the least time possible. Provision must be made for total drainage of water from sprinkler piping after each operation. Piping must be installed to pitch with auxiliary drain valves wherever water or condensate can accumulate or freeze. Heated valve houses are provided to enclose the dry-

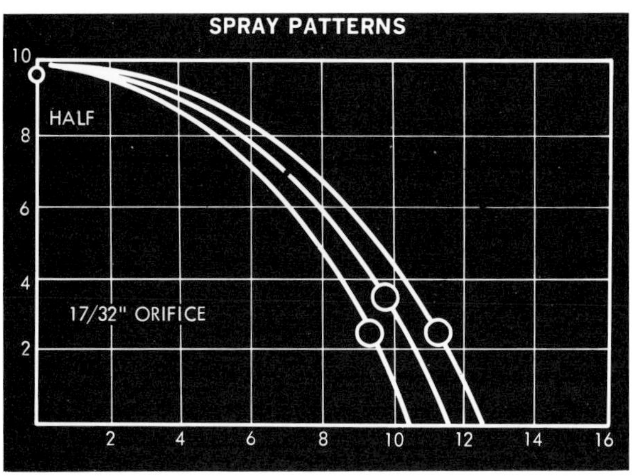

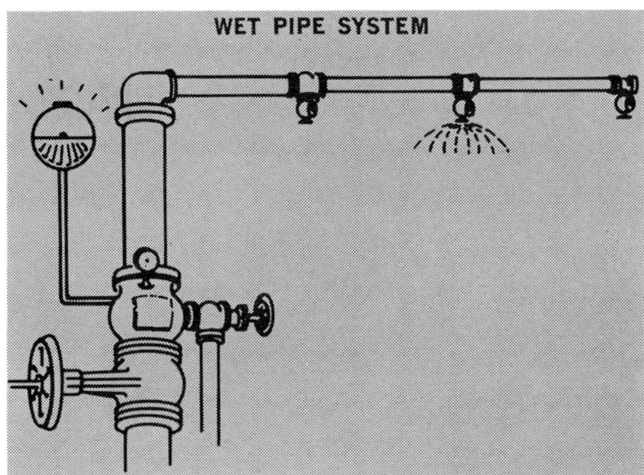

This system uses automatic sprinklers that are attached to a piping system containing water and connected to a water supply. Water discharges immediately from those sprinklers opened by a fire.

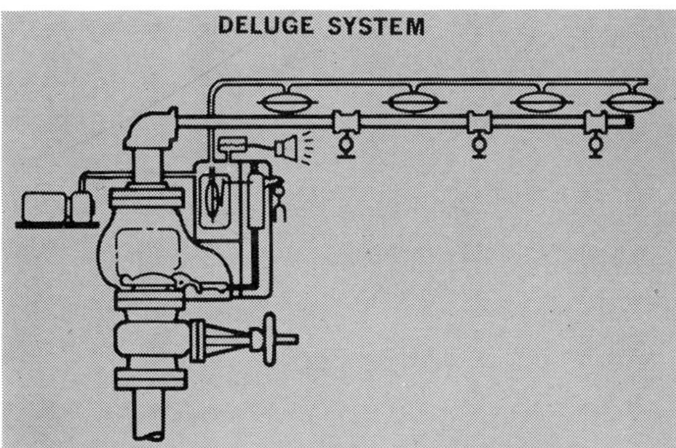

In the deluge system, all sprinklers are open, and normally there is no water in the piping. At the time of a fire, a supplementary detection system senses the fire and automatically opens a water control valve. This allows water to flow through the piping system to all sprinklers. Water then flows from all sprinklers in the system. The system shown uses H.A.D. (Heat Actuate Device) detectors and a drop-weight release to open the valve.

pipe valve and auxiliary equipment. Electric, steam, or hot water heat should be thermostatically controlled to limit temperatures between 40° and 70° F.

The preaction sprinkler system eliminates the possibility of premature discharge of water no matter the cause of sprinkler operation. Preaction features are appreciated by computer operators, department store owners, office building occupants, and managers of operations particularly vulnerable to delayed fire detection or to accidental water discharge. In this system, two events must take place before water will discharge from the sprinklers: First, a heat-sensitive detector must actuate the controlling valve to admit water to the dry piping and sound an early fire alarm; second, water will flow from the sprinklers only if fire extinguishers or other manual efforts at fire extinguishment are not effective.

The deluge system is operated by sensitive heat detectors; it simultaneously discharges water from open sprinklers connected to the overhead piping. Actuation is usually by heat-sensitive detectors. It is used, for example, where flammable liquid fires are expected to spread rapidly and envelop occupants, structural elements, storage tanks, and processing equipment. Rapid detection of fire and immediate wetting, blanketing, and cooling are necessary to prevent damage by heat, to prevent the spread of fire and to extinguish or control the burning. Airplane hangars, chemical-process and storage facilities, oil-filled power transformers and switchgear are examples of deluge-protected occupancies and equipment.

Specially Designed Systems

Modern exotic fuels and processes demand new approaches to fire protection systems. Tailor-made fire protection systems can be geared almost to anticipate the start of fire and to initiate fire safety operations with only a few milliseconds delay. Requirements for water densities or for unique extinguishing techniques are developed through large-scale experimentation. The cost of these one-of-a-kind research programs is small when compared to the potential cost of a single day's loss of production or difficult-to-replace process equipment.

Matching rapid detection to rapid actuation of the extinguishing process and to the proper extinguishing agent has produced fire protection systems that are a solution to previously hopeless problems. It is now possible to stop the rapid propagation of fire in such intimately mixed fuels and oxidizers as rocket propellants, pyrotechnics, dust suspensions and flammable gases. The rates of water application are several orders of magnitude above those used with ordinary flammables. Often, critical rates of application, below which additional hazards are created, are found (for example, water should not be applied to a magnesium fire unless the water rates can far exceed a threshold value).

Water-Flow Alarms

In addition to bringing a fire automatically under control, the sprinkler system rings a fire alarm and automatically notifies building occupants and the fire department that sprinklers are discharging water. The water-motor powered fire alarm is usually located on the outside of the building. Remote electrical signals can be transmitted over leased telephone lines directly to fire department headquarters.

Prompt response to these alarms with preplanned action limits both fire and water damage. Premature shut-off of sprinkler systems, however, can allow fires to propagate and to open additional sprinklers that will overtax the water supply when control valves are reopened. Failure of responsible personnel to be available to overhaul deeply hidden flames or to stop the needless discharge of water can contribute to additional damage.

The Mechanical Engineer

The mechanical engineer should consider the design and specification of fire protection systems early in the design stage, for early planning can lead to savings in design and construction costs.

Sprinkler piping support costs can be materially reduced with the early and strategic choice of roof and ceiling supporting techniques. Early study of fire protection water supply requirements will reduce the cost of both industrial and emergency equipment. Single integrated water systems can be strategically designed to satisfy all occupancy requirements.

Power supplies for pumping and signaling systems for emergency control are best studied at the same time as those for building lighting, heating, air conditioning, telephone and processes.

Structural elements, piping, ductwork, and equipment located in the ceiling should accommodate sprinkler piping. Planning of sprinkler piping, if done with a firm knowledge of location of other trades, will reduce added costs of relocating piping runs. Automatic sprinkler system designs and installations are specialized work. The skilled sprinkler fitter spends a number of years in apprentice training. The sprinkler

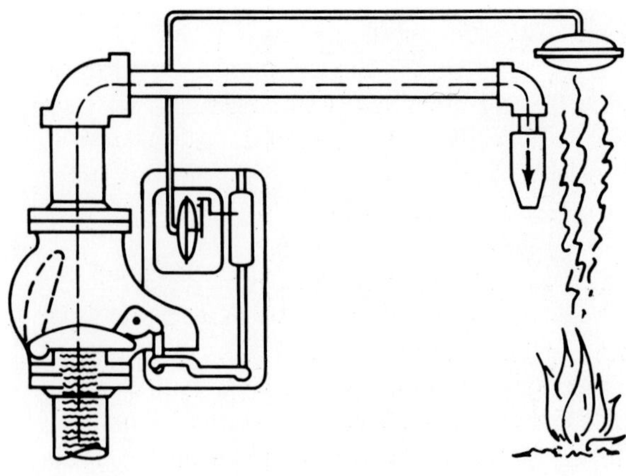

In an unprimed system, it takes time for the water to travel from the valve to the applicator.

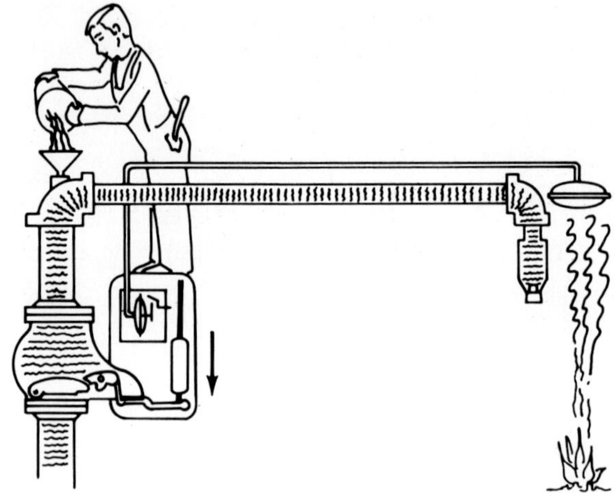

Time can be saved by priming the valve and using the priming water right up to the applicator where it is held by a rubber blow out plug.

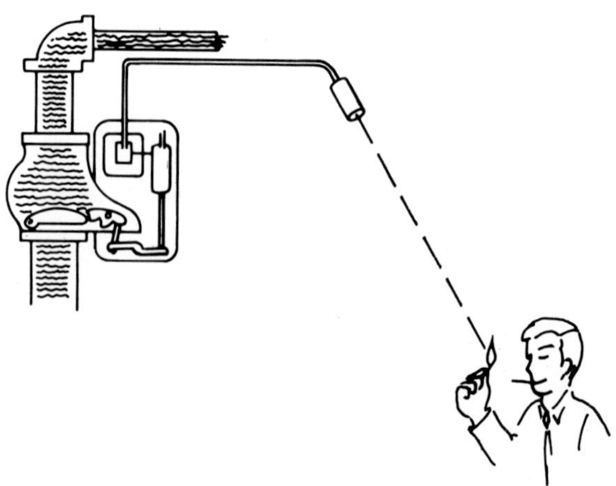

If rate-of-rise detection is still not fast enough, there is detecting equipment that can sense a fire electronically in milliseconds.

designer is experienced in the insurance requirements
and the special piping requirements of a fire protection
system. Specifications calling for design, installation,
and supervision by qualified automatic sprinkler contractors complement requirements for qualified contractors in other design and construction fields and
lead to better protection at lower costs.

Inclusion in the specifications of a sprinkler inspection and maintenance contract requirement will insure
years of uninterrupted dependable fire protection.
Sprinkler systems, unlike other facility utilities which
are used by the occupants every day, require periodic
test and inspection. This specialized emergency equipment can be maintained more dependably at less cost
by a qualified automatic sprinkler contractor.

The engineer who designs a sprinkling system should
understand the relationship between process, structural
materials, fire hazards of operation, and indemnity
for loss of operations due to any accident. He needs
to define the objectives of the fire protection systems
in meaningful terms; recommend research for new
processes; cite recognized standards in his specifications; anticipate the unusual situation and prepare
specifications for safeguards against that which should
not happen; and assist in evaluating proposals submitted by prospective contractors to meet the specifications.

References

1.   Reilly, Edward J., National Fire Protection
Assn., *Fire Journal,* Vol. 59, No. 1 (January, 1965),
p. 33.

2.   Bi-Monthly Fire Record, National Fire Protection
Assn., *Fire Journal,* Vol. 61, No. 3 (May, 1967), p. 56.

3.   Stevens, Richard E., National Fire Protection
Assn., *Fire Journal,* Vol. 60, No. 4 (July, 1966), p. 29.

4.   National Fire Protection Assn., Table 3–15.6.1,
"Standard for the Installation of Sprinkler Systems,"
NFPA No. 13 (1974).

5.   *Ibid.,* Table 3–15.5.

6.   "Automatic Sprinkler Performance Tables,
1970 Edition," *Fire Journal,* National Fire Protection
Assn., Vol. 64, No. 4 (1970), p. 37.

7.   *Op cit.,* p. 35.

8.   NFPA Standard for Installation of Sprinkler
Systems, NFPA No. 13 (1974).

9.   National Fire Protection Assn., Tables 3–4.1,
3–5.1, 3–6.1, "Standards for the Installation of Sprinkler Systems," NFPA No. 13 (1974).

Standpipe Systems

JAMES W. NOLAN

One of the major problems of tall buildings is design of adequate and reliable water supply systems to serve the permanent standpipe fire-fighting facilities.

Webster defines standpipe as a "high vertical pipe or reservoir for water, used to secure a uniform pressure." The definition seems adequate; however, as a fire protection engineer, I prefer the definition stated in Section 92-1 of the Chicago City Code: "An arrangement of piping installed in a building with outlets located in such a manner that water can be discharged in streams through hose attached to such hose outlets, for the purpose of extinguishing a fire and so protecting a building and its contents, with pumps, tanks and other equipment necessary to provide an adequate supply of water to the hose outlets."

NFPA Standard No. 14, 1974, Standard for the Installation of Standpipe and Hose Systems, groups standpipe systems into three classes of service by intended use in extinguishing fires:

Class I: For use by fire departments and those trained in using heavy fire streams (2 1/2-in. hose).

Class II: For use primarily by the building occupants until the arrival of the fire department (small hose).

Class III: For use by either fire departments and those trained in handling heavy hose streams or by the building occupants.

Where does the design engineer go for help in the design and development of a standpipe system? The answer has always been city, county or state codes, recognized model building codes or NFPA No. 14. Except perhaps for cities such as New York and Chica-go, none of these codes or standards covers the subject thoroughly, especially for buildings taller than 500 feet.

In the past, the designer of a high rise usually collaborated with municipal and insurance authorities in developing a satisfactory fire protection water supply system. From this collaboration local codes evolved for future buildings of similar height and structure, but they differed widely, not only in their content but also in their design philosophy. Such differences currently exist in both New York City and Chicago codes. Fortunately, NFPA No. 14 has clarified the problem somewhat and it is expected that future editions will cover the problem thoroughly.

Before proceeding with a standpipe design, the design engineer should always review the local fire prevention code requirements. If no code exists, he may find that an NFPA standard has been adopted by reference, as it has been in Illinois.

Number and Location

It would seem reasonable to establish the number of standpipes on the basis of floor area, if it were not for the problem of partitioning. To overcome that problem, the "length of hose" method has been established as a design criteria in NFPA No. 14. Simply stated, this standard requires that the number of hose stations (in effect, standpipes) shall be based on a hundred feet of hose as the maximum allowable length extended through undivided (or divided) rooms to within thirty feet of all portions of each story of the building. (OSHA, and some local codes base design on seventy-five feet of hose). In other words, the more partitions in the building, the more standpipes necessary to make sure that those hose streams can be directed into any room. Although exterior exposures should also be considered, standpipes are generally in or near stairwells.

To limit pressure, the system should be divided into zones. NFPA No. 14 normally limits the height of single-zone systems to 275 feet above grade. When hose outlets are equipped with fail-safe pressure regulating devices, installed in accord with Section 213 of NFPA 14–1974, base pressure of zones may be 250 psi. The zone limit varies from city to city. For example, the Chicago code requires two zones if a building is more than 260 feet above grade. New York, however, recommends zoning beyond 300 feet. In any event, the 275-foot limit will assure reasonable working pressures.

Additional independent zones limited to, say, twenty stories or 225 feet each should be added as the height of the building dictates. Each zone should have its own water supply, as well as independent Siamese connections. Thus, the system operating pressures of each zone will be within the design limits of centrifugal fire pumps, pipe and fittings, end-use products—such as hose valves, orifice discs, and hose—and the normal operating range of fire department pumpers.

Orifice discs or pressure-reducing devices must be installed on hose valves wherever pressure in the lower parts of a zone exceed 100 psi. Orifice sizes normally range from 3/8 inch to not more than 1/2 inch diameter. Pressure-reducing devices can be adjusted to give desired pressures at the hose nozzle and, if necessary, can be readjusted to give higher pressures.

Each standpipe should be sized for a minimum flow of 500 gpm. Where more than one standpipe is required, all common supply piping should be sized for a minimum flow of 500 gpm for the first standpipe and 250 gpm for each additional standpipe. NFPA No. 14, recommends four-inch pipe for standpipes up to one hundred feet high; six-inch pipe up to and including 200 feet in height; and, though not specifically stated, eight-inch pipe beyond 200 feet. When two or more zones are required, each zone should have separate and direct supply piping of not less than eight inches in diameter. High-level zones with two or more standpipes should have two direct supply pipes of at least eight inches in diameter.

Combined Systems

When system risers (vertical piping) are arranged to serve both two and a half-inch outlets for fire depart-

ment use and outlets for automatic sprinklers, they must be sized for both uses. Current requirements of NFPA No. 13 and 14 permit such risers to be sized by hydraulic calculations.

Water supply for combined systems (in completely sprinklered buildings) must be at least 500 gallons per minute for light hazard occupancy and at least 1,000 gpm for ordinary hazard occupancy. Also, in completely sprinklered buildings small hose (one and a half-inch) for use by occupants may be omitted.

Zone heights and pump sizes are based on the need to achieve 65 psi for hose use at the top of the zone and to avoid exceeding 175 psi at the lowest level of sprinkler piping (so extra-heavy fittings are not required on branch line piping).

Piping

Pipe used in standpipe systems will normally comply with ASTM Standard A–120 for black steel pipe or ASTM A–72 for wrought iron pipe. The test pressures of Schedule 40 Grade A standard-weight pipe, which has a tensile strength of 48,000 psi and a yield strength of 30,000 psi in four-inch, six-inch, and eight-inch diameters, is 1,200 to 1,300 psi. NFPA Standards permit standard-wall Schedule 40 pipe for pressures up to 300 psi. The test pressures of Grade B Schedule 40 standard-weight pipe, which has a tensile strength of 60,000 psi and a yield strength of 35,000 psi, are somewhat higher. The test pressures of the same size pipe Schedule 80 extra-strong is not much higher at 1,700 psi.

The fittings in the standpipe and connections should be of extra-heavy pattern where the pressures are or may be more than 175 psi. (Pump shut-off pressures under maximum supply pressure conditions give maximum pressure.) Fittings should be flanged for sizes greater than six inches. Joints may be welded, but permission should be obtained from the authority having jurisdiction. UL-listed 300-pound class ductile iron flanged fittings for working pressures up to 500 psi are commonly used in many standpipe systems. Approved expansion joints or flexible couplings that provide for linear expansion movement are also used in some parts of the country.

Gate and check valves should be of extra-heavy flanged pattern where the pressures are likely to be more than 175 psi. Although most standards state that approved straightway pattern-type check valves should be used, the silent-type check valve should be considered. Straightway pattern check valves on standpipe systems often create extreme water-hammer problems. For this reason, the nonslam check valve is preferred. This valve has a spring-operated clapper with a limited travel, and it may be installed either horizontally or vertically.

NFPA No. 14 recognizes the following water supplies:

1. Public waterworks system where pressure and discharge capacity are adequate
2. Automatic fire pumps (taking suction from city supply of adequate flow but inadequate pressure or from storage tanks or reservoirs)
3. Manually controlled fire pumps in combination with pressure tanks
4. Pressure tanks
5. Gravity tanks
6. Manually controlled fire pumps operated by remote control devices at each hose station

The water supply methods described here are limited to automatically controlled fire pumps or gravity tanks and to Class III service.

The water supply should be sufficient to provide 500 gpm for thirty minutes. Where more than one standpipe is required, the minimum supply should be 500 gpm for the first standpipe and 250 gpm for each additional standpipe and not exceed 2,500 gpm for at least thirty minutes. Thus, the size of automatic fire pumps would be as follows:

1 standpipe	500 gpm
2 standpipes	750 gpm
3 standpipes	1,000 gpm
4 standpipes	1,500 gpm
5 or more standpipes	2,000 gpm to 2,500 gpm maximum

This is a conservative evolution of the "250 gpm/standpipe" rule. The variation, however, reflects the sentiments of fire departments about what constitutes an acceptable water supply for each floor of a high rise.

NFPA recommendations say, "The [water] supply shall be sufficient to maintain a residual pressure of 65 lb/sq in. at the topmost outlet of each standpipe, including the roof outlet with 500 gpm flowing." Past recommendations and some current city codes, based on the use of half-inch solid-stream nozzles, required that the minimum pressure at the top hose outlet be 40 to 50 psi. Today, however, most fire departments use adjustable spray nozzles, which require more pressure than solid-stream nozzles. UL's Fire Protection Equipment List describes several one and a half-inch spray nozzles, all rated at different pressures. The NFPA committee, in establishing the higher 65-psi pressure base, considered the friction loss in a hundred feet of hose and the higher pressures required for the effective use of the adjustable spray nozzle. It was also assumed that UL would consider listing all nozzles at the specified figure.

NFPA No. 14 says, "An approved means of maintaining a positive pressure on all zones of standpipe systems shall be provided." Of the six acceptable water-supply methods that would satisfy this requirement, automatically controlled fire pumps seem to be the most desirable and commonly used. Gravity tanks are still applied in some cities, but high-rise designers usually resist using a gravity tank.

Three Workable Systems

Figure 10.1a (from NFPA No. 14) illustrates a single-zone system to a maximum height of 275 feet. The three standpipes necessitate the use of a 1,000-gpm fire pump connected to the city water supply. The top one hundred feet of each standpipe is four inches in diameter, the bottom hundred feet six inches; eight-inch piping in the basement serves each standpipe.

Figure 10.1b illustrates a two-zone system with a maximum height of 500 to 550 feet. The system includes two fire pumps, each rated 1,000 gpm. The first fire pump connected to the city water supply serves the low-level zone and supplies a high-pressure suction to the second fire pump, which serves the high-level zone. Both pumps and their respective controls are in the basement and are interwired to ensure that both pumps will start when pressure drops in the high-level zone. When there is a pressure drop in the low-level zone, only the first pump is started automatically. Two independent eight-inch express risers serve the high-level zone. Pipe sizing of each zone is determined as before. Each zone has its own fire department Siamese connection.

Figure 10.1c illustrates an alternative two-zone system. The first fire pump, serving the low-level zone, is in the basement, but the second fire pump is at the 275-foot level. This pump obtains its suction supply from a tank at the same level that is filled either from the low-level zone, as illustrated, or from a separate tank filling pump. The advantage of this system is that the second fire pump need not be specially constructed to accommodate the high-suction pressure of the first. The disadvantages: the need for 275 feet of protected power wiring to serve the high-level fire pump, at least two eight-inch express risers in the low-level zone to serve the high-level zone, and a separate eight-inch high-pressure express riser from the fire department's Siamese connection to the high-level zone.

Figure 10.1c is included because it appears in the 1974 edition of NFPA No. 14. The arrangement seems to be popular in some sections of the country.

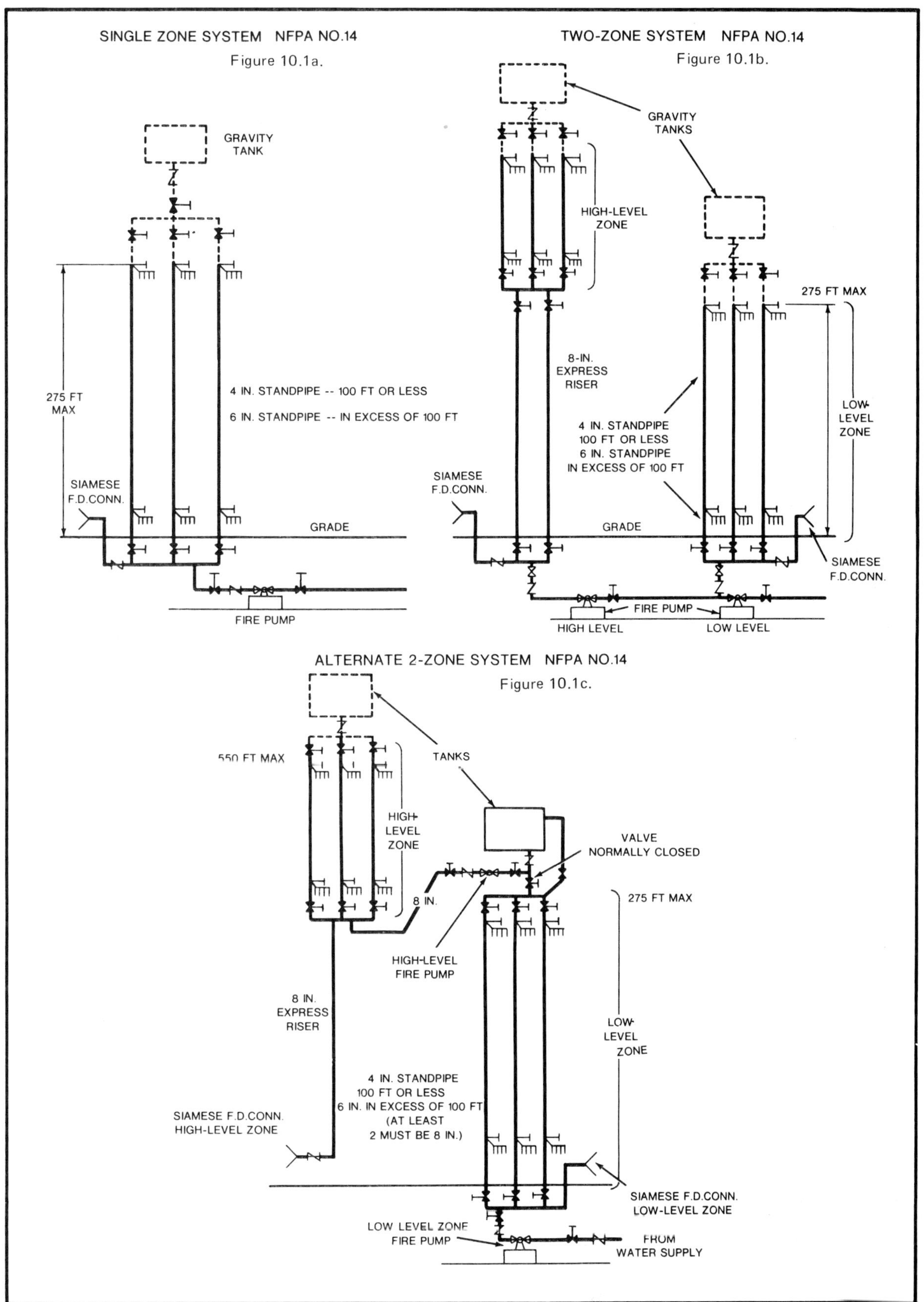

Figures are copyrighted by the National Fire Protection Assn. and are reproduced here with permission.

For the last several years, almost all fire department pumpers have been of the two-stage series-parallel high-pressure centrifugal type. Pumpers of this type must deliver 70 percent of rated capacity at not less than 200 psi design pressure, and 50 percent of rated capacity at not less than 250 psi design pressure. Consequently, a 1,000 gpm pumper is required to deliver 500 gpm at 250 psi design head. Residual city suction pressure would add to the pump design pressure, but in no case could the pump discharge pressure exceed the maximum relief-valve pressure setting of 300 psi. Moreover, fire department hose is required to pass an annual hydrostatic pressure test of 250 psi. The average 1,000 gpm city pumper, then, can be expected to deliver 500 gpm at an effective operating pressure up to an average height of 400 feet above grade.

It is at or near 400 feet that more automatic fire pumps with at least a thirty-minute stored water supply must be installed. A reliable secondary source of power supply—standby engines or engine/generator sets—should be considered mandatory. Most building codes now require this.

The IBM Building in Chicago has a three-zone system to a height of 700 feet (see Figure 10.2). This system includes three fire pumps, each rated 1,000 gpm. Like that of a two-zone system, the first fire pump is connected to the city water supply to serve the low-level zone and also supplies the second fire pump, serving the intermediate-level zone. Pumps and controls are in the basement. At 466 feet, a two-section 30,000-gallon reservoir provides a thirty-minute water supply

to the third fire pump, which serves the high-level zone. The diagram illustrates two methods of filling the reservoir; the first has an independent tank-filling riser served by two 500-gpm tank-fill pumps, and the second has a connection to the intermediate-level standpipes. If all else fails, the intermediate-level's fire department Siamese connection provides a third alternative. All provisions must be made, including a secondary source of power to the third fire pump, to ensure that an uninterrupted supply of water is available to serve the high-level zone.

The John Hancock Building in Chicago has a five-zone system to a height of 1,100 feet (Figure 10.3). Two fire pumps in the basement serve the lower part of the building, and three fire pumps at the 500-foot level serve the upper part. Because a 30,000-gallon domestic water reservoir was necessary for domestic use in the building, it was arranged that this reservoir feed down into a 30,000-gallon fire water reservoir, providing a combined storage capacity of 60,000 gallons of water available to upper-level fire pumps. The separate tank-filling line, a connection to the lower standpipe system and the fire department Siamese connection again would provide normal, as well as emergency, tank-filling methods.

The number of fire department Siamese connections for each zone depends on street exposure and building size. For every street exposure, most codes require at least one Siamese for each zone, regardless of the property's size. In some cities the number of Siamese connections required is increased for a street exposure.

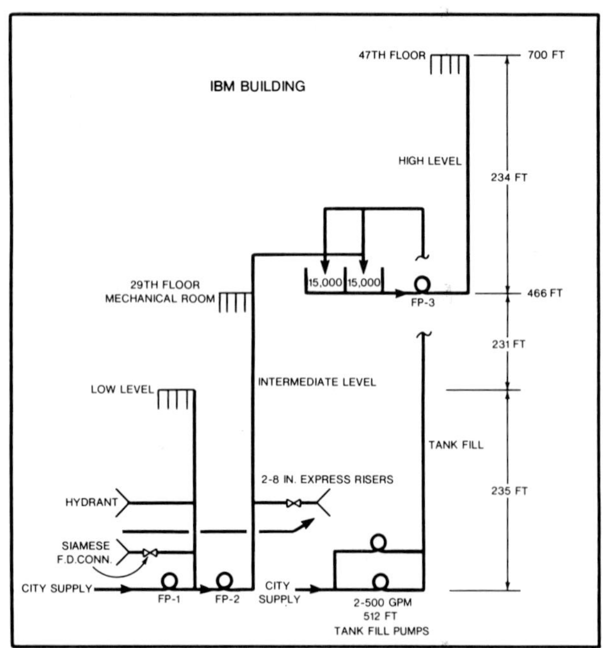

Figure 10.2.

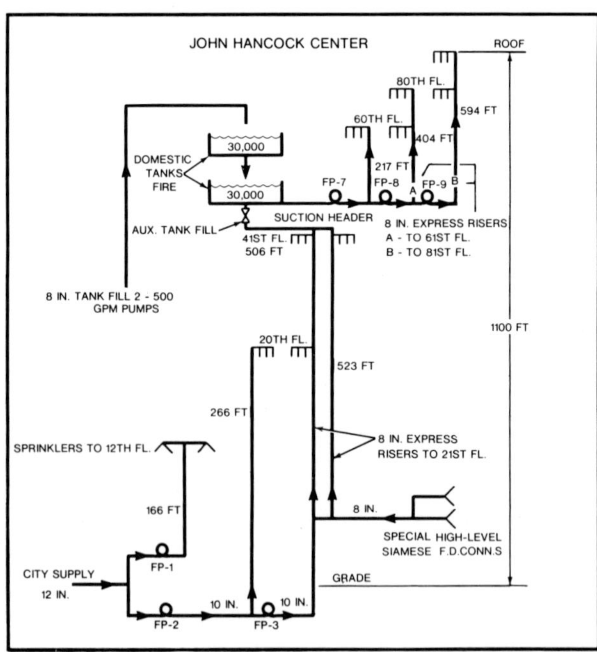

Figure 10.3.

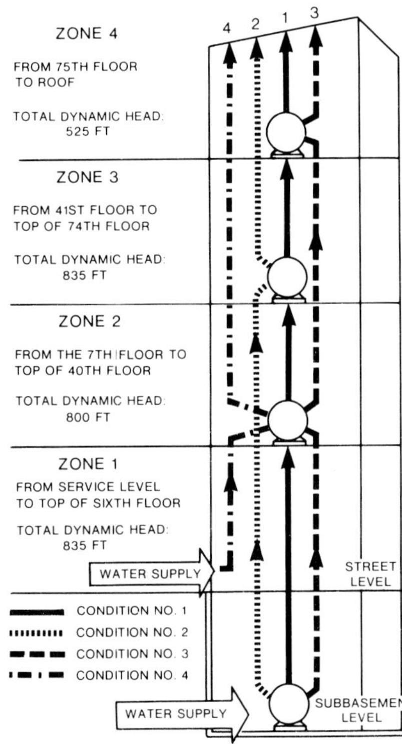

THE NEW YORK APPROACH

The new World Trade Center Building under construction in New York City is using a four-zone standpipe design that is a sophisticated variation of the alternate two-zone system. Each tower is divided into four zones, and each zone is served by a 750-gpm high-pressure fire pump. The Zone 1 fire pump is connected to the city water supply at the sub-basement level and supplies the Zone 2 fire pump on the 7th floor. The Zone 2 fire pump supplies the Zone 3 fire pump on the 41st floor. This pump, in turn, supplies the Zone 4 fire pump on the 75th floor, which provides 50 psi at the roof outlet.

Each pump is driven by a wound-rotor motor and controlled by a manual wound-rotor motor controller. Manual adjustment of the speed of each motor will operate the pumping equipment in either a series or bypass arrangement to provide for four emergency conditions. Condition 1 is a straight in-line series connection. In Condition 2, the first pump skips the second pump and goes to the third, skips the fourth and goes to the roof. In Condition 3, the first pump goes to the second, skips the third and goes to the fourth and then to the roof. In Condition 4, the Zone 2 pump is supplied with 300 psi, plus suction pressure supplied by the NYC super-pumper fire truck; the Zone 2 pump skips pumps 3 and 4 and goes directly to the roof. The system does not operate automatically.

For example, a building 250 feet long may need two Siamese connections per zone. Some authorities feel that a more conservative approach to the number of Siamese connections should be followed, although most fire departments would be satisfied with this exposure rule.

Buildings Under Construction

On March 27, 1968, a fire almost destroyed the top floors of the unfinished thirty-seven-story Equitable Life Assurance Building in Atlanta. No one was hurt, but damage was considerable. This fire in a high-rise building under construction emphasized the importance of installing a system of piping that will protect the building while it is being built. A temporary standpipe system should be either a part of the permanent installation or an independent temporary system, but in no case should the top hose outlets be more than two stories below the top-most construction level. In concrete construction this will be the top floor on which forms are being placed; in steel construction it will be the top floor on which the steel is being encased.

Hose valves two and a half-inches in diameter should be available on every floor for the fire department to use. Fire department Siamese connections should be identified and accessible. In buildings more than 500 feet tall, a permanent water supply system should be installed in the building and tested before the building reaches that height.

Water Spray Systems

CHARLES F. AVERILL

Water spray systems differ from the more common regular sprinkler system in their basic design and in the coverage and characteristics of the water discharged. Sprinkler systems provide protection for general areas, whereas water spray systems are designed for specific hazards—they may protect a general area or individual pieces of equipment. Water from sprinklers fall vertically in an umbrellalike pattern, but water spray nozzles can discharge in any direction and have a wide variety of discharge characteristics as to pattern and volume.

Water spray systems are most frequently used in the chemical, petrochemical, utility and refinery industries. There are also many fire hazards in general manufacturing that are protected with water spray systems. Typical hazards protected include aircraft hangars, alcohol storage, asphalt impregnating, cleaning plant equipment, drying ovens, escalators, stair wells, explosives manufacturing and storage, flammable liquids storage, flammable solids storage, fuel oil storage, hydraulic oil, lubricating oil, LPG storage, oil quenching baths, paint manufacturing and storage, petrochemical storage, record vaults, transformers, circuit breakers and turbine lubricating oil.

How Water Extinguishes

Water has been used to extinguish fire for as long as man has known about fire and is still the most commonly used agent. Probably more than 90 percent of all fires are put out with water. Since most fires are Class A (fires in ordinary combustibles, such as wood, paper, cloth), and water is the most effective extinguishant for this type of fire, it has become commonly associated with "ordinary fires."

Water is generally considered to be an improper extinguishing agent for Class B (flammable or combustible liquids or gases) or Class C (electrical) fires. It is in fact, however, a very effective extinguishant when properly applied.

It is usually assumed that water extinguishes fire by cooling—i.e., by reducing the temperature of the burning material below the point at which it can give off flammable vapors. (All materials, including wood and paper, burn in the vapor phase only.) Although this is generally true, other things are also happening. If fuel, oxygen and an adequate source of ignition are present simultaneously, fire results. This is known as the "fire triangle," but combustion is more complex than that. The fuel and oxygen (or some other oxidizer) must be in proper proportion (flammable range) to burn. A mixture outside this range is said to be too rich or too lean to burn. The upper and lower "flammable limits" of flammable liquids and gases are given in most reference books that list properties of combustible materials. (See Table 11.1.) Considering this property of combustion, it can be seen that if burning vapors can be diluted with nonflammable vapors to below the lower flammable limit (LFL), combustion will cease and the fire will go out.

When water is heated above 212° F., it, of course, turns to steam and, in so doing, absorbs a great deal of heat. Steam is a nonflammable vapor, and when it is mixed with flammable vapor it can reduce the mixture of flammable vapor and oxidizer to below the LFL.

Combustion is actually a chain reaction wherein flammable molecules collide with oxidizer molecules at a high enough temperature for them to combine. If enough nonflammable molecules are introduced to

Table 11.1. Flammable Limits of Some Common Chemicals

Name	Flammable Limits Percent by Volume		Name	Flammable Limits Percent by Volume	
	Lower	Upper		Lower	Upper
Acetic Acid (Glacial)	5.4	16.0	Fuel Oil No. 1	0.7	5.0
Acetic Anhydride	2.9	10.3	Gasoline	1.4	7.6
Acetone	2.6	12.8	Heptane—n	1.2	6.7
Acetylene	2.5	81.0	Hexane—n	1.1	7.5
Ammonia (Anhydrous)	16.0	25.0	Jet Fuels	0.6	3.7
Amyl Acetate—n	1.1	7.5	Methallyl Chloride	3.2	8.1
Benzol (Benzene)	1.3	7.1	Methyl Chloride	10.7	17.4
Butadiene—1,3	2.0	11.5	Methyl Ethyl Ketone	1.8	10.0
Butane—n	1.9	8.5	Naphthalene	0.9	5.9
Butyl Acetate—n	1.7	7.6	Pentane	1.5	7.8
Butyl Alcohol—n	1.4	11.2	Propane	2.2	9.5
Butyl Benzene—n	0.8	5.8	Propyl Alcohol—n	2.1	13.5
Carbon Disulfide	1.3	44.0	Propylene	2.0	11.1
Carbon Monoxide	12.5	74.0	Propylene Glycol	2.6	12.5
Dichloroethylene—1, 2	9.7	12.8	Propyl Ether—iso	1.4	21.0
Dimethyl Butane—2, 2	1.2	7.0	Vinyl Acetate	2.6	13.4
Ethyl Acetate	2.5	9.0	Vinyl Chloride	4.0	22.0
Ethylene	3.1	32.0	Xylene—o	1.0	6.0

reduce the number of collisions between flammable and oxidizer molecules below a critical point, the chain reaction is broken and combustion ceases.

Water does two things in putting out a fire: It cools the combustible to reduce the production of vapors, and it dilutes the vapors produced.

These basic principles of cooling and diluting apply whether the fire involves Class A or Class B materials. They show why water, applied in the proper form is very effective in extinguishing all Class A and some Class B materials.

Because most Class B materials burn more readily than Class A materials, more efficient cooling and diluting is required. This can be accomplished by applying the water in a spray form rather than in a straight stream. A third extinguishing effect water can have on some Class B materials is emulsification.

In the early 1930s in England it was found that if the surface of a pool of oil was bombarded with relatively large water droplets at high velocity an emulsion would form on the oil surface. The emulsion is unusual in that it is composed of droplets of oil surrounded

A water spray system in action at a typical chemical plant installation.

by a film of water. This results in continuous layers of water above the oil that cool, dilute and present a physical barrier reducing vapor emission. This emulsifying phenomenon can only be produced in oils having the viscosity of kerosene or a higher viscosity.

Two Basic Sprays

Though water in spray form is a proper and effective extinguishing agent for many Class B fires, the characteristics of the spray are important.

Water spray can generally be divided into two categories—fine spray and coarse spray. It is difficult to define these categories numerically since there is a wide range of drop sizes within the pattern of any spray nozzle. A recent paper by C. Yao [1] and A. S. Kalelkar, which deals with drop sizes produced by a standard sprinkler, establishes some principles that are valid for water spray systems as well as sprinkler systems. All extinguishment takes place at the interface of the base of the flame and the burning material; so the water spray drops must have sufficient mass and velocity to penetrate the flame zone.

Generally nozzles having a spray pattern formed by a jet of water impinging on a deflector produce a fine spray; nozzles having a spray pattern formed by an internal break-up of the jet produce a coarse spray. Fine spray is usually used for cooling and exposure protection. Coarse spray is used for extinguishment. Both type nozzles, however, frequently are used in the same system. Figures 11.1 and 11.2 show the two types of nozzles and the sprays they produce.

Where to Apply Sprays

Water spray systems are used for four purposes—singly or in combination. These are extinguishment, exposure protection, controlled burning and fire prevention.

Exposure protection is the prevention of damage from the heat of a nearby fire. This protection is particularly applicable to chemical and petrochemical plants. In some instances it is better in the long run to let a chemical fire burn itself out rather than extinguish it. This is particularly true when the source of fuel cannot be shut off and where controlled burning (with the exposures protected) results in less overall damage than if the fire were extinguished and unburned fuel allowed to spread over a large area where it could reignite. When an exposure fire can be extinguished in a reasonable time, exposure protection reduces the likelihood of spreading to other areas.

Controlled burning is accomplished by introducing water spray into the combustion zone. This reduces the intensity of the fire and the heat output; it is then easier to put the fire out with chemical extinguishing agents or to protect the exposures while the fire burns itself out.

Fire prevention in this context means the absorbing or diluting by means of air aspiration of flammable vapors before they become ignited, thereby preventing a fire.

System Design

Before any water spray system is specified the basic purpose of the protection should be outlined, since the design criteria is dependent on what the system is intended to accomplish. Such details as type of spray, rate of application, pressure required, location of nozzles, etc., vary, depending on the purpose of the system. The basic components of a system are the same, however. These are an adequate water supply, fire detection equipment, spray nozzles, a piping system, hangers and alarms.

The water supply's pressure, flow (gpm) and total volume must be sufficient to meet the intended purpose of the system. Pressures from 40 to 150 psi are suitable, depending on the purpose of the system. Flows from a few hundred to several thousand gpm may be required; the flow depends on the type and extent of the hazard. The required duration of the flow may be as little as two minutes where rapid extinguishment can be accomplished, but the duration may be two hours or more where exposure protection must be maintained for long periods.

It is obvious that dependability is of utmost importance in water spray systems, as in all fire protection systems. The service conditions are unique among all industrial systems. The protection system may be on standby for years. Since it is not essential to production, it tends to be forgotten; and unless special effort is made, maintenance is poor. It is essential, however, that this mechanical equipment, which may not have moved for years, function perfectly in a fire emergency. With this service condition in mind, the basic engineering principle that says the simpler the design and the less complex the mechanism, the more dependable the equipment obviously applies. Simple equipment, however, is limited in capability, and some sophistication is often necessary to accomplish the desired purpose. Therefore the specifying engineer must weigh the performance requirements against the dependability and maintenance characteristics. This is particularly true for fire detection equipment.

Figure 11.1. The external-breakup nozzle (inset) produces a fine spray for cooling and exposure protection.

Fire Detectors

Detectors are sensitive to heat, smoke, products of combustion, or light produced by fire. They are generally classified by the type of phenomena that triggers their operation—heat detectors, smoke detectors, etc. (The basic characteristics of fire detectors are explained in Chapters 20 and 21.) They are further classified by energy source—i.e., hydraulic, pneumatic or electric.

Heat detectors are most commonly used and are usually the simplest in construction. There are two basic types: fixed temperature (which operates at a predetermined temperature level) and rate of temperature rise. However, in the type of hazard where the anticipated fire would be large, such as in a major

Figure 11.2. The internal-breakup nozzle is used in systems needing a coarse spray for fire extinguishing.

flammable liquid spill, the difference in operating time between rate-of-rise and fixed temperature types becomes insignificant; therefore, selection should be based on other considerations.

Smoke or products-of-combustion detectors are very sensitive and usually operate at an earlier stage of the fire than a heat detector; however, they are more sophisticated in design and are dependent on electric power.

Light detectors are the fastest in operation. Their use makes possible the design of systems that have the capability of getting water to a fire in milliseconds. This type of detector should only be used where the hazard is such that extreme speed of operation is required or where conditions make other types of detectors unsuitable.

National Fire Protection Assn. standards and insurance requirements require that all detection systems

A water spray system protecting a power transformer station.

be supervised. This means that if the detection components of the system are damaged, a visual or audible signal is given. It is also required that where detectors use electrical energy, a standby power system, such as trickle-charged batteries, be provided.

Nozzles and Piping

Only spray nozzles that have been specifically designed and approved for fire protection use should be specified for water spray systems. There are a great number of different types of spray nozzles that vary in capacity, pattern, and discharge angles. Nozzles may be of open or closed orifice design to allow an entire system or only one nozzle to operate automatically. The final selection of the specific nozzles to be used in any given system, therefore, should be left to the fire protection engineer.

The piping system should be given special attention by the design engineer. The principal considerations are materials, location, sizing and hanging or support.

ASTM-A-120 pipe is the standard of the industry. Black pipe is used for indoor installations and galvanized outdoors. Standard-weight cast-iron screwed fittings are used where the piping system is not subject to high heat from a rapidly developing fire. Standard-weight malleable-iron or ductile-iron screwed fittings should be used in the fire area.

Generally speaking, cast-iron fittings are used indoors and malleable-iron or ductile iron outdoors; however, malleable-iron or ductile-iron fittings should be used in indoor systems where flammable liquids may be burning. Fittings used indoors are black; those used outdoors are galvanized.

Cast-iron flanged fittings are used in valve headers and large supply lines. Groove-type couplings and fittings may also be used. These fittings can also be used in the protected area if the fire exposure will not be severe before the piping can be filled with water. Welded construction is not required and is usually avoided because welding introduces a fire hazard and burns off galvanizing. The use of shop-fabricated welded headers and sections, galvanized after fabrication is acceptable and frequently used, particularly for such large, complicated pieces as multiple-valve headers and systems protecting power transformers.

Forged- or cast-steel fittings should not be used because the available sizes are limited and delivery times are long, which places an unnecessary burden on the designer. Special materials such as copper, brass and stainless steel are sometimes used for unique situations, such as corrosive atmospheres.

Location of the piping system is dictated by three considerations: the required location of the spray nozzles it feeds, the need for adequate hanging or support and the avoidance of interference with other systems or the building structure.

Since the spray nozzles direct the water on the fire, their location is of first importance. Nozzle position is somewhat limited by specified spacing rules, individual coverage patterns and effective range. However, once positioned, nozzles become a major factor in the location of piping.

The need for adequate pipe support is obvious, and piping is usually kept close to structural members so hangers can be easily attached. For example, in a grid system under a roof, the piping to which nozzles are attached is run at right angles to the purlins rather than parallel to them. This arrangement makes it possible to locate the piping as dictated by the nozzle locations and still use the purlins for hanging. This arrangement is shown in Figure 11.3.

The pipe must be sized so that each nozzle receives water at adequate flow and pressure and there is a hydraulic balance throughout the system, in order for the available water volume and pressure to be used most efficiently. This requires hydraulic calculation of the piping system. The principal factor to consider is the friction of the water flowing through the pipe. Velocity is unimportant and is usually not considered except in underground pipe, for which it is limited to 16 ft/sec to avoid shock at changes in direction.

In some systems where water supplies are marginal and water volume must be determined as accurately as possible, the effect of velocity pressure on the discharge from the end nozzles is taken into account. The generally accepted formula for calculating fire protection systems is the Hazen-Williams formula:

$$p = \frac{4.52}{C^{1.85}} \frac{Q^{1.85}}{d^{4.87}}$$

where:

Q = flow in gpm

d = pipe ID in inches

p = friction loss in psi/ft

C is a constant that is dependent on the roughness of the interior surface of the pipe. The accepted values of C for various types of pipe are these:

Cement-asbestos	140
Cast concrete	100
Unlined cast iron	100
Cement-lined cast iron	140
Tar-coated cast iron	100
Wrought iron	120
Bitumastic-enamel lined	
Cast iron or steel	140
Welded or seamless steel	120
Copper or brass	130

More information, example problems and a standardized calculation form are given in NFPA Standard No. 15, Water Spray Fixed Systems.[2] Anyone attempting these calculations should be familiar with and follow the procedure given in the standard.

Hangers

The hanging or support of a piping system for fire protection is an important design consideration. There are few if any other types of piping systems that have as high a safety factor built into the hangers. Hangers should be attached to structural members; hanging

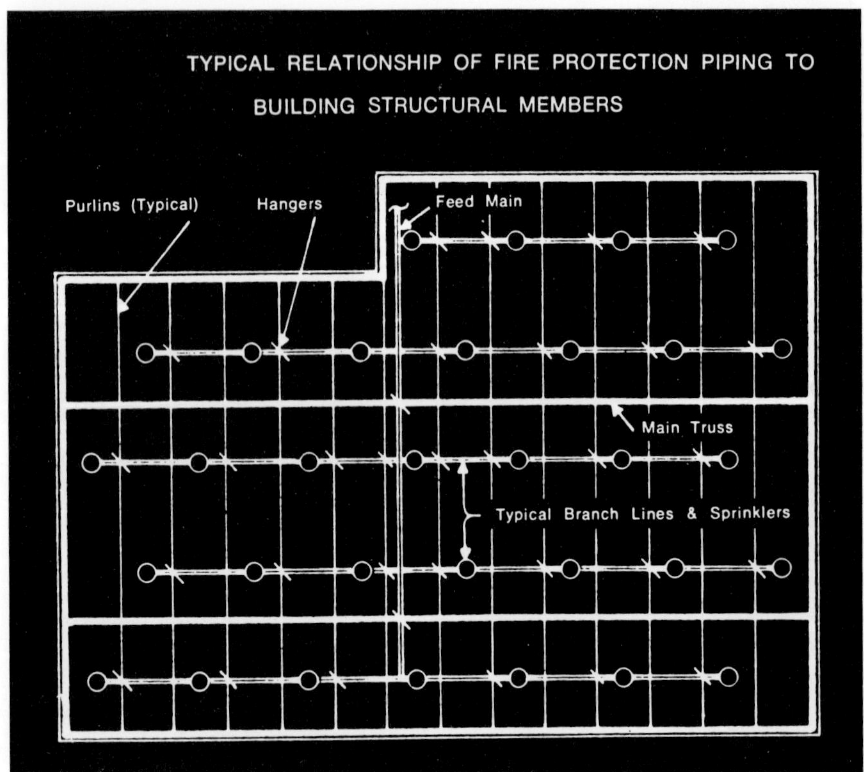

Figure 11.3.

from roof decking, floor sheathing, other piping, ductwork or process equipment should be avoided. Hangers
specifically designed and approved for fire protection
should be used. When standard hangers are inapplicable
and unique ones designed, the special hangers must
have the same rugged characteristics as standard hangers.
NFPA Standard No. 13, Installation of Sprinkler
Systems,[3] gives general information about hangers
for sprinkler systems. These hangers can frequently
be used in water spray systems; however, it is sometimes necessary to design special hangers, especially
for outdoor systems.

Two Kinds of Alarms

The basic alarms that should be provided for water
spray systems are trouble and fire. The trouble alarm
usually signals failure of the integrity of the detection
system or the closing of a main water-control valve.
One signal, usually a horn, is used for both trouble
conditions, since the trouble is quite obvious at the
valve location. The fire alarm, usually a bell, is actuated
by the opening of the automatic water-control valve
or by the signal that activates the valve.

The energy source for alarms can be electric, pneumatic, or hydraulic. The most dependable alarm is a
water-motor alarm, which needs only the system
water pressure for operation. However, this type of
alarm is more commonly used with sprinkler systems
than with water spray systems.

If the simple alarms described above are specified,
they should be part of the water spray system specifications. If, however, a complex system is required, the
alarm should be described in a separate section of the
specifications or included in the electrical section.
Complex alarm equipment should be installed by
companies specializing in this field.

References

1. Yao, C. and Kalelkar, A. S., "Effect of Drop
Size on Sprinkler Performance," Factory Mutual Research Center, Serial No. 18792.

2. Standard No. 15, "Water Spray Fixed Systems,"
National Fire Protection Assn.

3. Standard No. 13, "Installation of Sprinkler
Systems," National Fire Protection Assn.

Low-Expansion Foam Systems

D. N. MELDRUM and J. R. WILLIAMS, P.E.

Fire protection systems using low-expansion foam are important for the protection of flammable and combustible liquids. A properly engineered system produces a fluid, stable, and tenacious foam, which can be distributed evenly or spread over burning materials in sufficient quantity to smother, cool and thus extinguish fire. Foam makes a vapor-suppressing blanket that provides protection against reflash and reignition. For this reason, it is particularly suited for protecting storage tanks, aircraft hangars, dip tanks, drainboards and loading racks. It is also used for general fire-fighting purposes in refineries, chemical plants, storage terminals, tankers and barges, and for emergency fire-fighting and rescue operations where there is hazard to personnel from flammable liquid fires. Foam systems are classified as portable, semi-portable and fixed and may be manually or automatically actuated.

The development of fire-fighting foam paralleled the growth of the petroleum industry. The early foams were generated chemically by mixing a solution of alum with a solution of bicarbonate of soda and licorice-root extract. Since large volumes of solution and lead-lined vessels were required, storage was expensive. The development of a dry two-component and finally a dry single-component system followed. This was a marked improvement, but still, one pound of foam power was needed to generate one cubic foot of foam. Typically sized equipment required one to four fifty-pound cans of powder per minute. The logistics necessary to supply the foam chemical were more difficult than actually fighting the fire. Finally, the trend to larger tanks for the storage of petroleum products made chemical-foam systems impractical.

Mechanically generated foam from liquid concentrates was first developed in Germany in the 1930s. While these early concentrates produced a poor quality foam compared to today's mechanical foams, they provided a great advantage over the chemical foam powders because one gallon produced more than 300 gallons of foam. Also, the liquid concentrates could be introduced into a water system more easily than the powdered materials.

Foam systems can be characterized by line diagrams. For most foam systems the following pertains:

Proportioning:
Water + foam concentrate → foam solution
Generation:
Foam solution + air ⇋ foam

The first diagram represents what happens when the concentrate is introduced into a water supply or into a pipe line (using a proportioner). As shown, this function is irreversible. The second diagram shows what happens when the solution is made into foam with air added by a foam maker or foam nozzle. The fact that this part of the system is reversible shows that most foam wants to return to the foam solution state and that proper system design requires that foam be generated at an adequate rate.

Foam systems have two other requirements.

There must be a driving force or energy to move water and foam concentrate from a source or storage tank through piping to the point of discharge. A means of forming or directing the foam discharge onto a fire is also needed. This may be a nozzle (portable) or fixed-discharge device.

Modern mechanical foams are produced by mixing air with the foaming solution. Although compressed air can be used, economics limits practical fire equipment to using air inspirated by a venturi effect. The mixing needed for foam generation is done with a foam maker or foam nozzle. Expansion ranges of four to twelve (i.e., volume of foam generated divided by volume of solution used) are typical for foam designated "low-expansion." High-expansion foam (see Chapter 13) generally has expansions in excess of 100 to 1. Foams have been produced with expansion ratios between those of low-expansion and high-expansion foam. In some cases, these can be used as either high- or low-expansion foam, but no performance or system design parameters have yet been developed for them. The foams discussed in this chapter are limited to recognized and approved low-expansion foams.

The foam industry has developed a wide variety of foam concentrates to meet specific requirements. (These are discussed in more detail in the next section.) The following generalizations for use of low-expansion foams, however, set the application boundaries:

The hazard must be a liquid.

The hazard must be below its boiling point at the ambient condition of pressure and temperature.

Care must be taken if the bulk temperature of the liquid is higher than 100° C.

The hazard must not be unduly destructive to the foam selected.

The hazard must not be water-reactive.

The fire must be a surface fire. Three-dimensional fires cannot be extinguished by foam unless the hazard has a relatively high flashpoint and can be cooled by the water in the foam.

The quantity of foam applied and rate of application must be matched to the size, fuel and character of the fire.

With low-expansion foam the primary mechanism of extinguishment is smothering. Theoretically, if the hazard were completely covered with an infinitely thin blanket of noncombustible material, the fire would be extinguished. In practical circumstances, however, the minimum foam depth is about 1/4 inch; the average depth is three or more inches. Flammable liquids—those with flashpoints below 140° F.—are extinguished primarily by this mechanism.

The second mechanism, cooling, plays some part in the control of combustible liquid (those with flashpoints higher than 140° F.) fires. Here, if the surface of the liquid can be cooled below the flashpoint, the fire will be extinguished. Heat transfer occurring through evaporation of the water in the foam and by heat absorption of the bulk foam temperature can reduce the fuel surface temperature if the rate of burning is reduced.

Characteristics of Foams

The ideal fire-fighting foam would have all of the following characteristics:

1. Hold its water content for a long period to permit formation of a vapor-tight blanket over the entire liquid surface
2. Flow rapidly and easily over the fuel surface
3. Protect against flashback indefinitely

It would also be able to cling to vertical surfaces; resistant to fuel attack; easy to clean up; nontoxic or irritating; biodegradable; highly concentrated; easy to proportion and apply; noncorrosive; economical; usable under all expected ambient conditions.

Because of technical limitations, no foam extinguishing system exists with these qualities. A wide number of foaming compounds are available, however, that are formulated to be most suitable for the conditions expected.

Types of Foam Concentrates

Regular protein foam is prepared from purified animal or vegetable protein, stabilizers (including iron salts), necessary solvents and an industrial germicide to protect against bacterial decomposition in storage. This material has been available since the early 1940s. It is produced in 6 percent and 3 percent concentrates (6 percent = 6 parts concentrate plus 94 parts water) in the United States and in a range of concentrations—primarily 4 percent—in foreign countries. (See Table 12.1 on characteristics of low-expansion foam.)

Regular foam (3 and 6 percent) is suitable for use on nonpolar hydrocarbon fuels. The primary advantages of this foam are resistance to burnback, good extinguishment and low cost. It is also biodegradable, nontoxic, noncorrosive, easy to clean up and can be stored at 35° to 150° F. Special formulations are available for use at low temperatures. Regular foams have a low-temperature limit of 35° F. They are also available with low-temperature ratings of 20° F.,

Table 12.1. Characteristics of Low-Expansion Foams

Foam concentrate	Percent concentration	Typical storage temp. limits	Storage container construction	Application	Hazards	Approx. cost ratio/gal * (most concentrates)
Regular protein	3 or 6	20 F to 120 F	Mild steel, brass, 304 SS, PVC, fiberglass, neoprene	Type I or II	Hydrocarbons	1
Low-temperature	3 or 6	–20 F or –40 F to 120 F	Mild steel, brass, 304 SS, PVC, fiberglass, neoprene	Type I or II	Hydrocarbons	1.5
Fluoroprotein	3 or 6	20 F to 120 F	Mild steel, brass, 304 SS, PVC, fiberglass, neoprene	Type I or II subsurface	Hydrocarbons	2
Protein-based polar-solvent	6	20 F to 120 F	Mild steel, 304 SS. Do not use brass in construction contact	Type I only	Polar solvents	2
Polymer-based polar-solvent	20 + 3	35 F to 80 F	Catalyst use same as regular foam concentrate, 304 SS, fiberglass, PVC only	Type I or II	Polar solvents	1
Aqueous film-forming foam	6	35 F to 120 F	304 SS, fiberglass, Teflon-covered diaphragms only	Type I or II	Hydrocarbons	4.5

*Installed cost ratios will vary considerably from this list. Each job must be evaluated individually for meaningful comparison.

–20° F. and –40° F. These foams must be used where potential storage temperatures are lower than for regular foams.

Fluoroprotein foams are a recent improvement of regular protein foam. The addition of a fluorocarbon surfactant to protein foam retains the good characteristics of the original foam and provides the additional benefits of improved dry-chemical compatibility and increased fire resistance. It also makes subsurface foam application practical. Like regular foam, fluoroprotein foam is available in 3 percent and 6 percent grades. It can, therefore, be substituted in existing systems without changing the system.

The superiority of fluoroprotein foam has been demonstrated in several recent fires in which rapid extinguishment was achieved when regular protein foam was having difficulty. The theory behind the advantages of fluoroprotein is that the highly fluorinated surfactant orients in the bubble wall with the fluorine-containing "tails" outside the bubble. Since the fluorinated "tail" is also incompatible with the fuel, it tends to make the bubble shed the fuel and, therefore, protects the integrity of the foam. This is particularly important in subsurface foam application. As with regular protein foams, fluoroprotein foam is for use on nonpolar hazards.

Protein-based, polar-solvent-type foams (alcohol foams) are special foams for fire control of polar solvents. Regular protein, fluoroprotein and surfactant foams are rapidly destroyed by solvent attack of polar solvents. A simple way to determine if a polar-solvent foam is needed is to check the chemical composition of the fuel. If oxygen, nitrogen or sulfur is present, the hazard probably requires a polar-solvent type or other special foam. In such cases, the foam manufacturer should be contacted for recommendations.

Two types of protein-based polar-solvent foams are available. One is an alkaline material with a pH of approximately 11. In this material, the active system is based on zinc tetra-amine soap. The second uses a solublized aluminum soap. This slightly acid material has a pH of approximately 6. These types are, of course, incompatible and should never be mixed.

Protein-based polar-solvent foams work this way: An insoluble soap is formed when the concentrate is diluted with water. If the soap is distributed evenly in the bubble wall, a stabilized system is formed. As some of the foam drains, a diluted layer is formed on the surface of the fuel. This layer helps to protect the remaining stabilized form.

Limitations caused by this mechanism are extremely important. If a polar-solvent fire is to be extinguished, the following must be remembered.

The foam must be applied gently. NFPA No. 11 recognizes this application as Type I. If the diluted layer is disturbed by too vigorous application, the foam will continue to break down until the bulk of the liquid is diluted to a less active state. When the active soap is precipitated it will agglomerate if too much time is allowed between mixing and foam formation. This period is called transit time. Foams of this type have acceptable transit times of from less than five seconds to about two minutes under average conditions. With warmer water, the allowable transit time decreases; conversely, it increases when colder water is used.

Polymer-type, polar-solvent foams were developed to extinguish fires in aliphatic (fatty) amines. These particular compounds were extremely destructive to existing protein-based polar-solvent foams. The polymer-type foams proved tough enough to be applied by Type II or less-gentle application. For the first time, foam could be practically applied in depth to polar solvents by means of a nozzle. This was of immediate benefit to the shipping industry, because gentle application on tankers is impossible. This meant that large chemical tankers could be built with a practical means for deck fire protection.

Polymer-type foam is generated by aeration of a mixture containing 20 parts prepolymer solution, 3 parts catalyst and 77 parts water. The solution so formed has no precipitation problem; therefore, transit time is not a factor. Performance is extremely good on water-soluble compounds. Water-semisoluble and some water-insoluble compounds may require higher application rates.

Synthetic-hydrocarbon surfactant foams (wetting agents) have been on the market for more than twenty-five years, but they have not had great success. These foams are compatible with hydrocarbons and are fast-draining compared to protein foams. Hydrocarbon compatibility causes wicking of the fuel, and rapid drainage weakens the foam blanket. The result is poor sealability, poor fire-resistance characteristics or resistance to reflash.

Some of these agents have UL-listings as wetting agents or penetrants for Class A fires; a few are UL-listed for Class B fires. The specifying engineer must be careful to discern which materials fall in this class and to avoid these materials except in special cases. Where wetting agent foams are required, UL-listed high-expansion foams used through low-expansion foam-generating equipment generally do an excellent job.

Surfactant foams are usually corrosive to mild steel and should not be held in contact with brass for extended periods of time. Most engineering plastics are satisfactory.

Aqueous-film-forming foams are products of modern chemistry. They are not found in nature and have unusual properties. By tailoring molecules to have low spreading coefficients and high resistance to hydrocarbons, aqueous-film-forming foams (AFFF) were created.

AFFF has the fastest extinguishment rate of any known foam. This quality is best applied in such areas as crash rescue for airplanes and helicopters and for petroleum processing spill fires. Because of its rapid drainage rate and relative lack of resistance to burnback, AFFF may be less effective than protein foam for use on tank fires. As with other synthetic concentrates, particular care in choosing materials of construction for handling these concentrates is required.

Equipment for Foam Formation

Making foam requires two steps: first, mixing foam liquid concentrate with water to form a solution and, second, mixing the solution with air to form foam. In foam terminology, the first is "proportioning" and the second is "foam generation."

Foam liquid concentrate can be proportioned into the water stream by premixing, using energy from the water stream or by using an external power source.

Premixing has limited value because limited volumes are used and the liquid must be pressurized from atmospheric to the discharge pressure. Hydrant pressure cannot be used. Also, stored solutions degrade much more rapidly when they are diluted than when they are held as a concentrate.

Protein-type polar-solvent foams cannot be used as a premix. Premixed solutions of both protein and AFFF should be used or discarded after three to six months.

When the quantity of concentrate is not too large, and when proportioning pressure drop can be tolerated, energy from the water stream is used to induct the concentrate.

Venturi or line proportioners (see Figure 12.1) are the lowest cost proportioning devices. The operating range, however, is limited to one flow rate/pressure combination, and foam generation devices must be carefully sized to match the proportioner. In operation, water flows through the venturi creating a vacuum area. Foam liquid is introduced at this point by means of a pickup tube from the storage container. The venturi proportioner will meter accurately at a given flow rate/pressure point. Since proportioning rates vary with flows, these devices have a finite range of usage (usually expressed as a pressure limit). For practical fire-fighting purposes these devices operate satisfactorily at rated pressure ± 50 percent. The maximum recovered pressure, including friction loss and static head, is about 65 percent of the inlet pressure.

In pressure proportioners (see Figure 12.2), a small amount of the flowing water volumetrically displaces foam liquid into the main water stream. The working pressure of the vessel must, of course, be above the maximum static water pressure encountered in the system. Water is allowed to enter the foam tank from

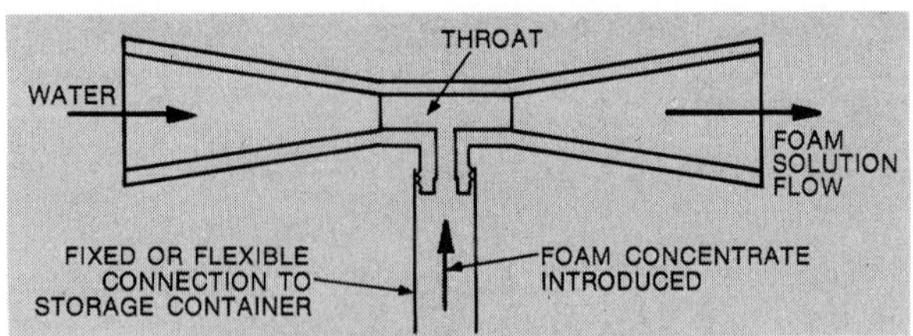

Figure 12.1. Line Proportioner (Venturi Type)

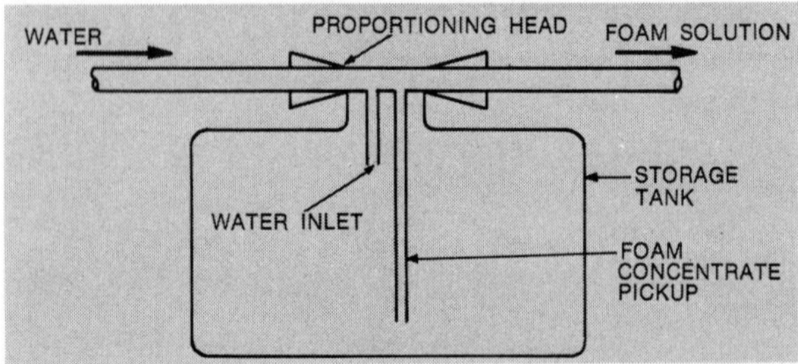

Figure 12.2. Pressure Proportioner

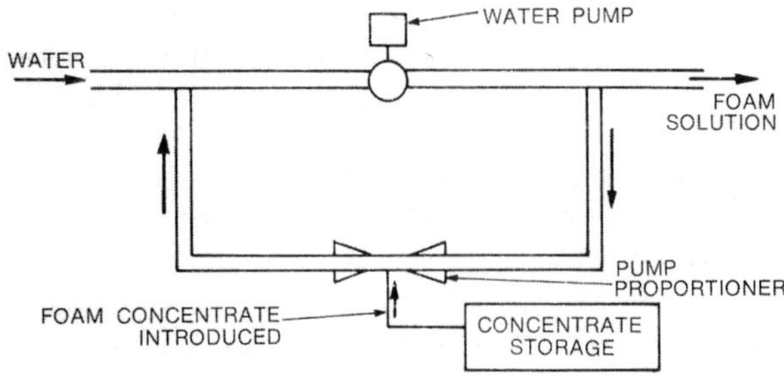

Figure 12.3. Around-the-Pump Proportioner

the main stream with as little friction loss as possible, while pressure in the main stream is dropped about 10 percent by means of an orifice. Liquid in the tank, essentially at the initial pressure, is metered into the low-pressure area by a second orifice. Advantages of this system are low pressure drop, automatic proportioning over a range of flows and pressures and freedom from external power. Disadvantages are long refill time (tank must be drained) and an economic limit on size.

There are two types of tanks available. The first contains a flexible bladder that separates the water and foam concentrate. While this is considerably more expensive, it allows for complete recovery of the liquid if a partial tank is used. The second type depends on specific gravity and viscosity difference to keep the water and liquid separate. This is entirely satisfactory for most fire-fighting liquids, and it is widely used.

It should be noted that pressure-proportioning systems can become inoperable or lose efficiency if there is water leakage through the main shutoff valves. This can be easily prevented by installing a ball drip valve between the supply valve and the main tank.

Around-the-pump proportioners (see Figure 12.3) are used where water is to be pumped on-site. Basically, the unit is a venturi proportioner that takes a portion of the main water flow, inducts a large percentage of foam liquid and injects the flow back into the suction side of the pump. This type of proportioning has the same limitations as venturi proportioners except recovered pressure is almost 100 percent. (Note: This is effected at some loss in pump capacity.) Around-the-

pump proportioners are generally very economical; but since they induct a fixed amount of concentrate independent of main line flow, they should not be used in a variable-flow system.

Water motor proportioners consist of a gear pump for the liquid driven by a water motor that uses pressure from the main water stream. Flows are limited to a fixed range.

Proportioning with external power is the simplest way to proportion. Compressed air or a gas cylinder is used to move liquid from a pressurized tank through an orifice into the water stream. This is done in small devices for specialty applications, but it is generally limited to less than ten gallons of liquid.

Direct pumping (see Figure 12.4) can be used when the foam liquid flow is fixed. Normally, a gear, turbine or piston pump is chosen in order to have positive displacement and metered discharge. Direct pumping through an orifice can be done with a centrifugal pump if the water pressure is fairly constant and if the proper pump curve is selected. Direct pumping is the lowest cost pumped system, but it is applicable only to fixed-flow systems.

Indirect orificing can be used when more than one flow is desired. In this type of system, the water-supply pressure must be adjusted to a fixed value. For the lowest flow the largest bypass orifice is opened, thereby bypassing a large amount of the foam liquid concentrate and injecting the proper amount. Since more solution is required when other devices are used, the orifice size

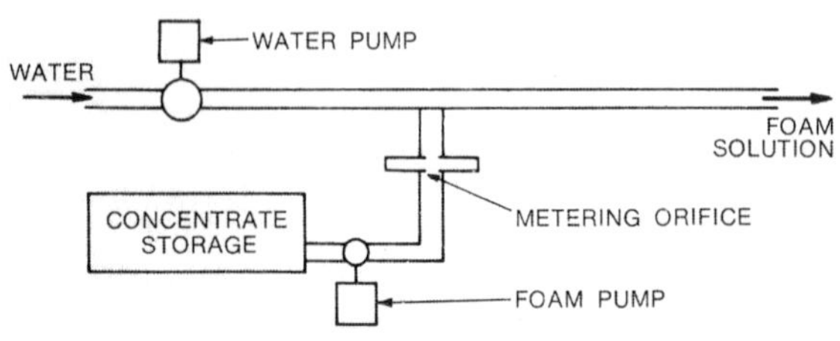

Figure 12.4. Direct Pumping

must be carefully selected to bypass the required amount of foam liquid concentrate. The largest number of different flows for bypass systems is usually three or four.

Indirect orificing can be extended to an infinite number of flows within a given range by using a variable orifice (valve) and measuring the water and foam liquid flows. This is called flow-indication proportioning. The greatest limitation here is the need for an attendant who normally observes rotometers and control flows. When the foam system is operated, water flow is indicated by a meter and the operator adjusts the liquid flow to obtain the desired foam liquid injection. Flow meters can be sized for a fixed injection percentage. Other percentages can be selected with the help of a single conversion chart.

Flow indication can be automated by using electronic or pneumatic flow sensors and a feedback loop to control either pump speed or the bypass valve. This is the most accurate system but is little used because of its high initial cost and its need for continual maintenance. The only possiblity for failure is a broken diaphragm, which is unlikely if proper materials are used. An auxiliary manual bypass of this valve is always included on balanced-pressure systems so the system can be operated under emergency conditions. Under practical conditions, the automatic range in which a balanced-pressure system will operate is approximately six times the lowest flow. Mechanical considerations (such as the unequal area of top and bottom diaphragms resulting from the shape of the stem diameter, for example) cause balanced-pressure systems to proportion slightly high at low flows. For example, in a nominal 3 percent system, the proportioning could be expected to be 2.9 percent at 600 gpm and 3.5 percent at 100 gpm.

In balanced-pressure proportioning (see Figure 12.5), automatic proportioning is provided through a mechanical system. Two orifices discharge water and foam concentrate into a common pipe. By adjusting the area of the orifices to a particular ratio, the percent injection can be adjusted if inlet pressures are equal. Equal pressures are maintained by using a diaphragm-operated valve on the foam liquid bypass line. Water pressure is applied to the top of the diaphragm and foam liquid to the bottom. If water pressure increases above the liquid the bypass valve will be closed, allowing less liquid to be bypassed and increasing the foam liquid pressure. The device is very simple in principle.

Devices That Generate Foam

Nozzles and fixed foam makers are devices that mix air with the proportioned foam solution. All low-expansion fire-fighting foam is generated by inspiration of air using a venturi and the foaming solution (except when a foam pump is used). Foam makers are divided into two classes: low back-pressure, which immediately discharge foam essentially at atmospheric pressure, and high back-pressure, which discharge the foam into a pipe for transfer or into the bottom of a tank for subsurface application.

Many types of low back-pressure foam makers are available. Each has a specific use and can best be recommended by the manufacturer. These types include foam nozzles ranging in capacity from 30 gpm to 4,000 gpm (solution rates are always specified for capacity), chambers for use on storage tanks, foam makers for dike walls, special units for dip tanks, foam water sprinklers and overhead foam sprays.

High back-pressure foam makers are similar to low back-pressure in that they also induct air by venturi action of the solution. They differ in that they are designed to recover up to 30 percent of the applied inlet pressure. When fluoroprotein foam is used, high back-pressure foam makers make practical the subsurface, or base, injection of foam into hydrocarbon tanks.

When selecting foam generating devices, be sure that they are designed and approved for the intended use.

Foam Application Requirements

How foam is applied to the burning surface of a hazard is very important to the success of extinguishment. The more gently the foam is applied the better the chance of success. Three types of methods are recognized by NFPA Standards:

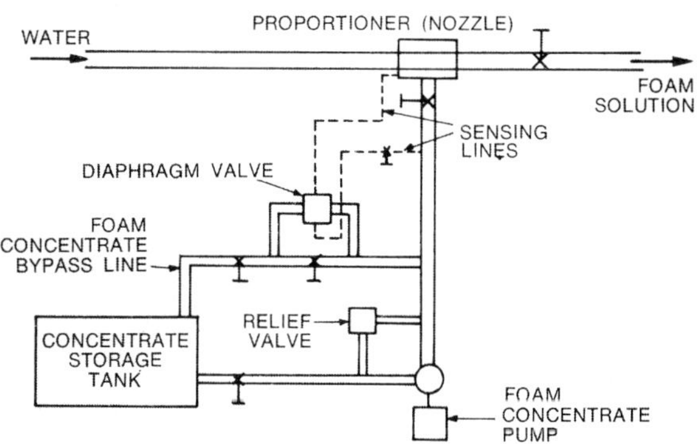

Figure 12.5. Balanced-Pressure Proportioning

Type I discharge outlet: An approved discharge outlet which will conduct and deliver foam gently onto the liquid surface without submergence of the foam or agitation of the surface.

Type II discharge outlet: An approved discharge outlet which does not deliver foam gently onto the liquid surface, but is designed to lessen submergence of the foam and agitation of the surface.

Subsurface foam injection: Discharge of foam into a storage tank from an outlet at the tank bottom or below the liquid surface.

With protein-based, polar-solvent foam, Type I application is a necessity. Even when Type I application is not required, its use will reduce operating time.

Table 12.2. Foam Discharge Times—NFPA 11-30

	Type of foam	
For tanks containing liquid hydrocarbons	Discharge Type I	Outlet Type II
Lubricating oils, dry viscous residuum (more than 50 sec Saybolt-Fural at 122 F); dry fuel oils, etc., with flashpoint above 200 F	15 min	25 min
Kerosene, light furnace oils, diesel fuels, etc. with flashpoint from 100 F to 200 F	20 min	30 min
Gasoline, naphtha, benzol and similar liquids with flashpoint below 100 F	30 min	55 min
Crude petroleum	30 min	55 min

Facts About System Design

The initial consideration in selecting a fire-protection system is the nature of the hazard. If flammable or combustible liquids are the hazard, foam should be considered. At this point, the authority having jurisdiction or the manufacturer should be contacted for recommendations on which type of foam to use. If the chemical composition and size of the hazard are known, a foam type and rate can be selected. This dictates the water demand.

Water supply is a limiting factor in many installations. Unless water can be supplied at the required rate and pressure, the foam system may be inoperable. Pumps can be used to increase pressure if the volume is sufficient. For most foam systems, application rates range between 0.1 and 0.6 gpm/sq ft of hazard, depending on the hazard and application method. It is important to have good quality water for foam generation. Potable water is always satisfactory. Difficulties may be encountered if the water contains antifoams, oil and, in some cases, high concentrations of surfactants.

Quantities of foam concentrate required by NFPA Standard 11 vary considerably depending on the type of hazard. For indoor dip tanks, for example, where the fuel level remains constant, the requirement is a two minute supply of foam at an application rate of 0.16 gpm/sq ft. For other indoor tanks, the supply must be 20 minutes. Outdoor hazards require a minimum of 0.10 gpm/sq ft. (See Table 12.2 on foam discharge times.)

Fluid-flow calculations must be made for all foam systems. Fortunately, foam solution behaves like water when flowing through pipe and, therefore, simplified calculations and flow tables can be used. Once the flow required at each foam discharge and the water pressure/volume at the source is known, calculations can be made. The system is balanced either by use of flow losses in pipe for fixed-capacity devices (i.e., nozzles, overhead sprays, foam water sprinklers), or by changing the foam-maker orifice in chambers, for example.

A wide variety of foam systems can be devised to protect flammable liquids. Typical indoor uses include dip tank system, floor flooding, overhead sprays or foam water sprinklers, one-man nozzle stations and oscillating monitors in hangars. Outdoor uses include storage tank systems with fixed chambers, diked area protection, loading rack systems and also oscillating monitors. Portable equipment is designed for use either indoors or outdoors, but it requires additional time to arrive and be put in operation.

Here are some important miscellaneous points for those specifying foam systems to remember:

In the U.S. and Canada, check for approval (Underwriters Laboratories, Factory Mutual Laboratories or Underwriters Laboratories of Canada) of equipment and agent.

The storage life of foam liquids is usually five to ten years depending on the type. Contamination, dilution and heat are the major factors affecting foam liquid life. If in doubt have the material tested by the supplier.

Design of foam systems—particularly for fuels other than hydrocarbons—requires special knowledge of the proprietary foam. Manufacturers should be consulted for the design of all systems.

For a comprehensive bibliography on foam, consult pages 17–45 of the *Fire Protection Handbook,* 13th Edition.

References

Fire Protection Handbook, 13th Edition, ed. George Tryon, NFPA, Boston.

NFPA Standard 11, *Foam Extinguishing Systems,* 1970, NFPA, Boston.

D. N. Meldrum, J. R. Williams, C. J. Conway, "Storage Life and Utility of Mechanical Fire Fighting Foam Liquids," *Fire Technology,* Volume 1 and 2, May, 1965.

J. R. Williams. "A New Foam for Polar Solvents," *NFPA Quarterly,* July, 1964.

Engineering Manual, National Foam System, West Chester, Pa.

NFPA Standard 11-b, *Synthetic Foam and Combined Agent Systems,* 1973, NFPA, Boston.

High-Expansion Foam Systems

D. N. MELDRUM

High-expansion foam has been praised, maligned, tested, pseudo-tested, and demonstrated successfully and unsuccessfully. Because of its impressive and uncommon appearance (the agent is sometimes referred to as "happy foam" and tends to flow and bounce over and around obstructions), high-expansion foam has been the delight of fire department shows and has been used by the television medium as a run-away shampoo, as a means for dissolving people and weapons (it won't really) and, occasionally, for fighting fires. Yet, in little more than ten years, high-expansion foam has progressed from a curiosity to a valuable fire-control agent.

Water, expanded by air and stabilized by surface-active agents as a large-celled foam, was introduced as a fire-fighter in the early 1950s. Unlike conventional low-expansion foam, which controls flammable liquid fires primarily by surface coverage, high-expansion foam, with its relatively high air-to-solution ratio, is basically a filling or volumetric agent. By definition, a foam in the expansion-ratio range of 100 to 1,000 to 1 is a high-expansion foam. (The expansion ratio is the ratio of volume of foam to the volume of solution from which it was formed; it is the reciprocal of the density of the foam.)

Although primarily composed of air, high-expansion foam depends on water contained in the foam cell or bubble interstices for its capability to control and extinguish fires. Originally, the foam was developed by the Safety in Mines Research Establishment of Buxton, England, to control mine fires. In established mine fires, access for effective reach of water nozzle streams is nearly impossible. However, researchers found that by wetting a porous net with a solution containing a foam-stabilizing, surface-active agent, bubbles formed by air passing through the wetted net would travel as a "plug" in the direction of air flow. Hence, through control of mine ventilation, water, transported by the foam plug, could be directed to the fire.

This type of fire control or extinguishment resulted from two actions: cooling by heat absorption, and dilution of the oxygen concentration in the confined mine entry as the water in the foam was converted to steam.

In 1956, Will B. Jamison, of the Pittsburgh Consolidation Coal Co., pioneered similar research in the United States. Following studies by the U.S. Bureau of Mines, several United States and foreign manufacturers began developing the new foam commercially. High-expansion foam systems are now approved for use in combination with sprinklers, as well as for spot-protection in areas where they augment existing fire protection systems. Used alone, high-expansion foam is an effective fire control agent, but mop-up or overhaul may be necessary to extinguish deep-seated fires completely.

Several types of high-expansion foam-generating equipment are now available as both portable units and fixed, pre-engineered systems. They all produce foam in the same way—air is passed over a porous contact medium (either aspirated or forced by fan) on which an aqueous solution of stabilized surface-active agent is sprayed. All foams are the same, regardless of how they are generated, when produced from un-

High-expansion foam generators mounted on the roof of an industrial plant. They make up a total-flooding system.

Photo credit: Walter Kidde & Co., Belleville, N.J.

contaminated air at a given expansion. The only variable is their water-retention ability, which thus becomes the most important criteria for comparison.

Early high-expansion foam concentrates were simple alkyl or aryl sulfonate salt solutions. However, these foams lost their water rapidly, and thus lost their ability to flow and carry enough water to fires burning anywhere but immediately in front of the foam generator. They have since been modified and stabilized.

A standard method for testing water-retention ability has been developed and is now described in NFPA 11A.

How It Works

As mentioned before, the primary extinguisher in high-expansion foam is the water. Efficiency of the agent and the mechanism of extinguishment vary with foam stability, its expansion, and the nature of the fuel and the physical configuration of the fuel. Although the combinations are almost limitless, here are some general guidelines:

Flammable Liquids

Extinguishment of flammable liquids (flash point below 140° F.) is primarily by smothering, i.e., combustion-sustaining air is excluded. Effectiveness of the foam depends on its stability, which in turn depends on its expansion, the chemical nature of the foaming agent, the temperature and nature of the water used for foam generation, and the uniformity of the bubble structure. No synthetic surfactant-based foam of any expansion has yet proved so effective in suppressing flammable-liquid vapor-release as the low expansion conventional foams highly stabilized by proteinacious foam agents of high molecular weight. (Vapor-release suppression is valuable in securing an extinguished flammable liquid fire.) High-expansion foam will not extinguish a running or falling (three-dimensional) flammable liquid fire. High-expansion foam is not recommended for use on polar solvents or water-soluble flammable liquids because surface activity or dilution will cause the foam to collapse. For hazard areas where flammable-liquid spills may be anticipated, low-expansion foam should be used for total extinguishment.

Combustible Liquids

Extinguishment of combustible liquids (flash point of 140° F. or higher) by high-expansion foam is primarily a combined effect of cooling and oxygen dilution. In tests or demonstrations, the foam appears to have a smothering effect, like that of low-expansion foam. Actually, as the foam advances, the foam front is destroyed. The concomitant conversion to steam absorbs heat from the fuel reducing its vaporization rate, and the steam itself dilutes the remaining vapors. It acts the same on combustible liquids as it does on flammable liquids: the more stable the foam, the greater its efficiency in terms of water requirements.

A fire-detection device triggers the generators and foam discharges into the protected
area.

Photo credit: Walter Kidde & Co.,
Belleville, N.J.

Class A Combustibles

Extinguishment is by oxygen dilution from steam
formation, cooling, and quenching. When generated
into a confined space in enough volume, high-expansion
foam may starve the fire by physically impeding the
access of oxygen. Oxygen carried within the foam is
sufficiently diluted by steam and will not support
the combustion reaction. The water in foam having
an expansion ratio of 1,000:1, for example, can provide
enough steam to reduce the oxygen concentration of
the resultant air/steam mixture to about 7.5 percent
by volume. For this to happen, however, there must be
enough foam; generally, the fuel must be totally submerged. If the fuel provides its own oxygen, extinguishment may not occur.

Foam will cool only those surfaces it touches. As

The high-expansion foam fills the plant
volume and extinguishes the fire. At this
point, the foam is 21.5 ft deep.

Photo credit: Walter Kidde & Co.,
Belleville, N.J.

it converts to steam, fresh foam is exposed (if foam generation is continuous) and further cooling results. The ability of high-expansion foam to penetrate low-density combustibles, such as rolled tissue paper or foamed plastic, has received limited qualitative attention. In the absence of adequate test data, its efficiency on such materials should be assumed to be low. While high-expansion foam can *control* fires in these materials, overhaul by hose streams is usually required for total extinguishment.

High-expansion foam, because of its whiteness and volumetric nature, can seem to submerge completely, and apparently extinguish, fires involving several types of combustible materials, while the fire continues to burn quietly underneath. This is particularly true with low-density, Class A combustibles, but the phenomenon has also occurred with flammable petroleum fuels. Test pool fire areas totally covered by high-expansion foam—or so it seemed to observers—actually had burning vapors beneath the foam, supporting the blanket on heated air. In such cases, overhaul, which can be dangerous, becomes a necessity.

Where to Use High-Expansion Foam

Because of its generally volumetric nature and since it extinguishes most fuels by oxygen dilution, high-expansion foam is best suited for confined spaces. The lack of confining walls and weather limit its use outdoors.

Most high-expansion foam generators can operate effectively at temperatures as low as –20° F., if the water is kept flowing, but the foam, which is relatively light in weight, is easily disrupted by wind or dry sunny weather. Its use in fixed systems is limited largely to areas where ventilation can be controlled and where foam flow can be directed. (The foam will flow in the direction of least resistance.)

Until tests prove otherwise, this foam should not be used on water-reactive chemicals or metals or on energized, unenclosed electric equipment.

Even though the most economical arrangement and installation of foam-generating equipment and the simplest system arrangement can be achieved by using air from within the hazard area for foam generation, current test data dictates against use of such air.

Research conducted by manufacturers and, more recently, by the U.S. Naval Applied Science Laboratory and the Naval Radiological Laboratory, show that excess concentrations of products such as carbon dioxide, carbon monoxide or nitrogen do not severely affect high-expansion foam. However, nitrogen oxide and nitrogen dioxide reduce foam volume and certain organic vapors (e.g., acetone) and hydrogen chloride cause severe structural breakdown. Moreover, the effective foam-generation rate decreases logarithmically as inlet-air temperature increases.

Considering these findings and the myriad unknown pyrolysis products yet untested, the NFPA Committee on Foam and insuring agencies require that air for foam generation *be drawn from sources outside the*

When the protected area is totally flooded, some foam escapes through openings in vents, windows or doors.

Photo credit. Walter Kidde & Co , Belleville, N.J.

hazard area and away from external sources of heat
or smoke.

Application systems for high-expansion foam fall
into three general categories: portable units, total
flooding systems and spot-protection systems.

Portable Applications

Portable units are available from several manufacturers in capacities ranging from 500 to 5,000 cfm of
foam. Trailer and truck-mounted units in capacities
up to about 13,000 cfm also are available. Portable
units have been successfully used primarily by fire departments to force water into inherently inaccessible
spaces, such as basements, truck trailers and attics.
These units may be used to augment other agent systems
and to purge spaces of heat and smoke by volumetric
displacement.

Total Flooding Systems

To date, most total flooding systems have been
used as adjuncts to sprinkler systems. The state of the
art of high-expansion foam used by itself is not ad-
vanced enough to indicate its advantages—either economic or performance—over sprinkler systems for
protecting Class A combustibles.

One drawback to total, high-expansion flooding
systems is their ability to cause damage. Totally sub-
merging of a warehouse in high-expansion foam in
controlling a small fire, for example, can cause a greater
dollar loss than the damage caused by localized sprinklers. Although high-expansion foam alone can achieve
fire control, it may not provide adequate fire protection.
From an insurance and protection viewpoint, however,
when combined with water sprinklers, high-expansion
foam is acceptable and sometimes necessary for totally
extinguishing particular hazards.

Rolled paper, for example, can be effectively protected by water sprinklers at storage heights up to
twenty feet. As stacking height increases, however,
sprinkler discharge becomes less effective. When high-

expansion foam is used as an adjunct, sprinkler effectiveness increases. In fact, tests have proved that for
high-piled, rolled paper stock this combination of
agents provides the best protection. Similar findings
apply to rubber tire storage. (The Rack Storage of
Materials Committee of the NFPA is currently developing a standard for sprinkler and high-expansion foam
systems based on fire tests now in progress.)

Total flooding systems are generally used to protect
rooms, storage areas, warehouses, laboratories and
buildings containing Class A or Class B combustibles
or both. Total flooding presupposes that an enclosure
surrounds the hazard, permitting volumetric filling
with foam. Leakage from the enclosure must be prevented—so windows, doors and other openings must
close automatically. Because the foam is opaque,
adequate warning of system actuation must be given
to allow personnel to get out.

Spot Protection

High-expansion foam can be applied locally to
specific hazards in a structure. This application calls
for a fixed foam-generating device complete with a
piped-in supply of foam concentrate and water. The
system is arranged to discharge foam onto the hazard.
For example, such systems may be required to provide
added protection for increased-hazard occupancy in
a previously sprinklered area. They are generally best
suited to protecting flat surfaces where combustible
liquids may spill. Several types of self-contained systems
are suitable for this service. Their applications include
use in small-hazard areas such as laboratories, paint
shops, unattended confined areas and where piped-in
water or electric supplies are not readily available.

A typical self-contained system is shown in Figure 13.1.

High-Expansion System Design

The requirements for pipe and fittings and methods
for hydraulic calculations for a high-expansion system
are identical to those for a sprinkler system, but there
the similarity ends. New concepts and components
required in high-expansion systems include a method
to determine the total amount of water required;
equipment for storing and proportioning foam concentrate into the water; venting equipment; and high-expansion generators.

The volume of foam to be generated is determined
by the kind and size of hazard, the contents, and the
possiblity that sprinklers will also be used. Tests have
shown that the minimum total depth of foam should
be at least 1.1 times the height of the highest hazard
and at least two feet over that hazard. For flammable

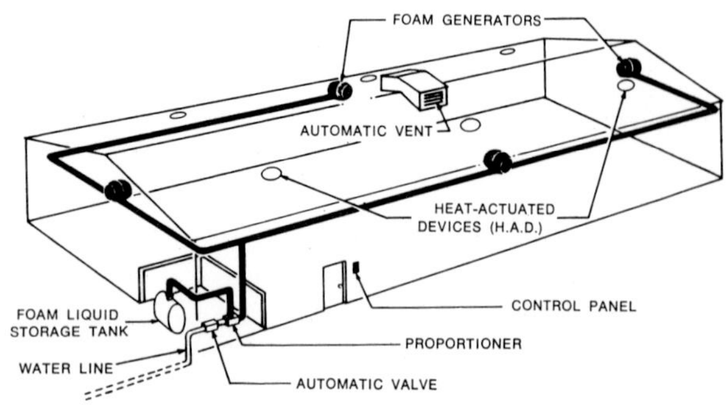

MAJOR COMPONENTS OF A TOTAL FLOODING SYSTEM

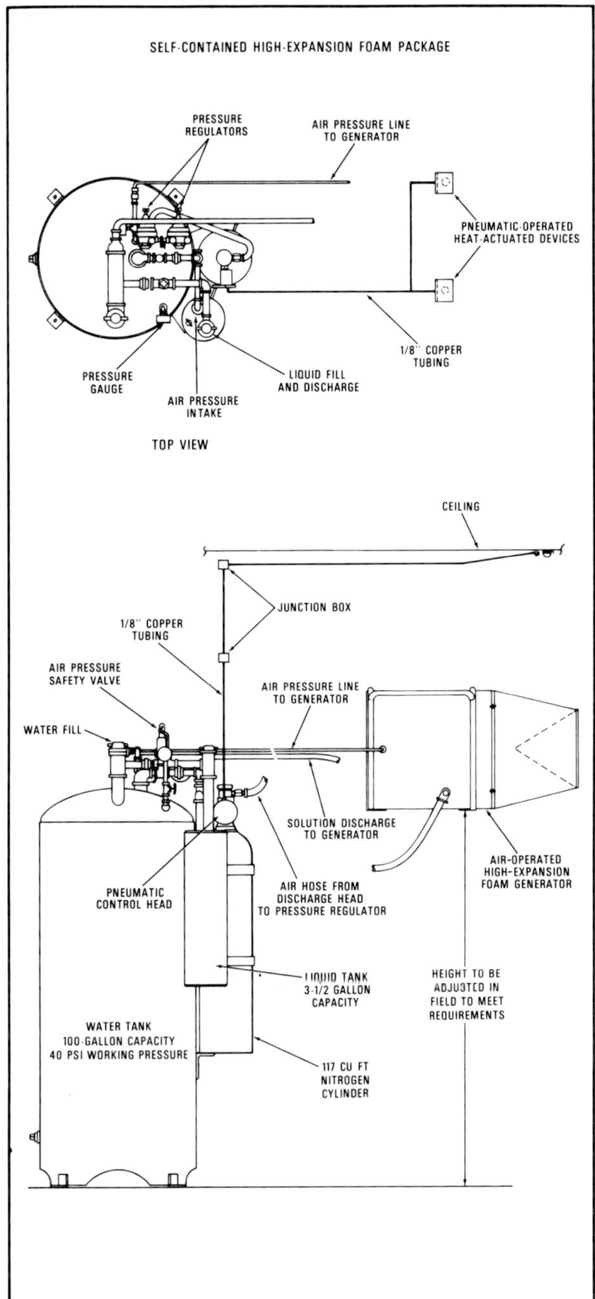

Figure 13.1.

or combustible liquids (each should be tested under simulated-hazard conditions), at least three feet of high-expansion foam should cover the hazard. NFPA 11A contains recommended maximum times to achieve submergence of the hazard. For design purposes, this is the product of the required foam depth and the floor area of the space protected.

Table 13.1 shows the major design requirements and a typical calculation problem.

The location of foam generators is up to the engineer, and no strict rules can cover all variations. The equip-ment can be bulky and complex, and it is, therefore, subject to physical damage and malfunction if not properly maintained. No specific standards have been developed, except that requiring an even buildup of foam in the protected area. The general trend is toward requiring several small foam generators to protect a given area, rather than a few large generators. Increasing the number of generators will increase system costs but permit a built-in margin of safety and, theoretically, increase total-system reliability if one or more genera-tors fail. Also, distribution of foam over the hazard is more uniform. Standards require that foam-generating equipment be placed as close as possible to the hazard being protected and that the generators be protected from fire exposure or physical damage.

Types of Foam Generators

Here are the types of foam-generating equipment generally available, with comments on their advantages and disadvantages:

1. *Air-aspirating.* Advantage: simplicity, no moving parts. Disadvantage: relatively low foam-expansion ratios require more water. These units are usually used as portable equipment.
2. *Fan-driven by electric motor.* Advantage: constant speed; highest foam volumes and force against back-pressure. Disadvantage: require reliable power source.
3. *Fan-driven by water motor.* Advantage: no electric power required. Disadvantage: fan speed, foam output vary with water pressure; foam volume gener-ally lower for given water flow, as compared with other means.
4. *Fan-driven by air motor.* Advantage: no electric power, constant speed if enough air and pressure available. Disadvantage: reliable air source required.

Unlike most water-discharge or low-expansion foam-discharge devices, high-expansion foam generators, at their current stage of development, are prone to mechanical failure. This is a result of their complexity and light construction. Their efficiency depends on the interrelationship of water pressure, fan speed, net configuration and integrity and foam-concentrate proportioning. Periodic—at least annual—tests are recommended. If stock or the building may be damaged by the water in the foam, a means of diverting test-produced foam is necessary.

The production of good-quality foam depends on the proper percentage of foam concentrate in the water stream. This will vary between 1.5 and 2 percent. Several approved methods of proportioning for foam concentrate are described in NFPA 11A. The most

Table 13.1. Maximum Submergence Time for High Expansion Foam Measured from Start of Foam Discharge* (Minutes)

Hazard	Light or unprotected steel construction		Heavy or protected or fire resistive construction	
	Sprinklered	Not sprinklered	Sprinklered	Not sprinklered
Flammable liquids (flash points below 140 F)**	3	2	5	3
Combustible liquids (flash points of 140 F and above)**	4	3	6	4
Low-density combustibles (foam rubber, foam plastics, rolled tissue or crepe paper)	4	3***	6	4***
High-density combustibles (rolled paper—kraft or coated—banded)	7	5***	8	6***
High-density combustibles (rolled paper—kraft or coated—unbanded)	5	4***	6	5***
Rubber tires	7	5***	8	6***
Combustibles in cartons, bags, fiber drums	7	5***	8	6***

*Based on a maximum of 30 sec delay between fire detection and start of foam discharge. Any delays in excess of 30 sec shall be deducted from the submergence times.

**Polar solvents not included in this table. Flammable liquids having boiling points less than 100 F may require high application rates. Where use of high expansion foam is contemplated on these materials, the foam equipment supplier shall substantiate suitability for the intended use.

***These submergence times may not be directly applicable to high-piled storage above 15 ft, or where fire spread through combustible contents is very rapid.

common ways to proportion high-expansion foam concentrate are shown in Figure 13.2.

How to Calculate Foam Volume

Using the table showing maximum submergence time as a guide for minimum requirements, the minimum rate-of-discharge (total-generator capacity) can be calculated as follows:

$$R = \left(\frac{V}{T} + R_s\right) C_N C_L$$

Where:

R = rate of discharge (cfm)

V = submergence volume (cu ft)

T = submergence time (min)

R_s = rate of foam breakdown by sprinklers (cfm)

C_N = compensation for normal foam shrinkage

C_L = compensation for leakage

The factor (R_s) for compensation for breakdown by sprinkler discharge is determined by test. If specific test data is not available, the following formula must be used:

$$R_s = s \times Q$$

Where:

s = foam breakdown in cfm/gpm of sprinkler discharge. (s is 10 cfm/gpm)

Q = estimated total-discharge from maximum number of sprinklers expected to operate (gpm)

The factor (C_N) for compensation for normal foam shrinkage is 1.15. (This is an empirical factor based on average reduction in foam quantity from solution drainage, fire, surface wetting, stock absorbency, etc.)

The factor for compensation for loss of foam owing to leakage around doors and windows and through unclosable openings (C_L), is determined by the design engineer. Obviously, this factor cannot be less than 1 even for a structure completely tight below the design filling depth. This factor could be as high as 1.2 for a building with all openings normally closed depending on foam-expansion ratio, sprinkler operation and foam depth.

Here is a sample calculation of total generation capacity:

Given: Building size = 100 ft × 200 ft × 30 ft high. Building construction is light bar joist, Class I steel-deck roof, adequately vented. Masonry walls with all openings closable. Sprinkler protection is a wet system, 10 ft × 10 ft spacing, 0.25 gpm/sq ft

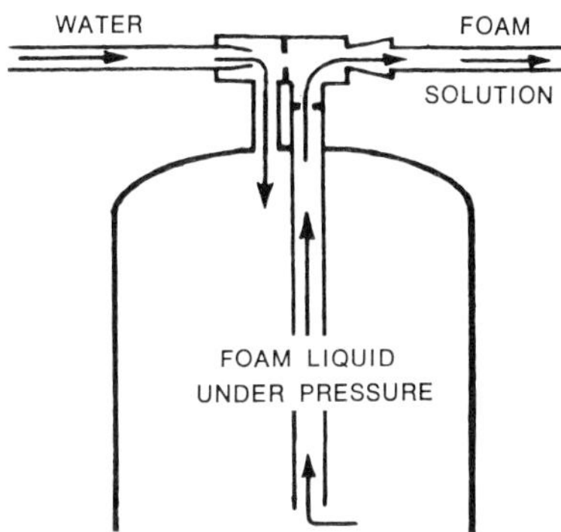

Water under pressure flows through the line proportioner venturi where a vacuum is created, inducing high-expansion foam liquid from the atmospheric storage tank into the water supply. The foam solution passes to the high-expansion foam generator where the foam is produced. This is a simple, inexpensive proportioning method when water supply pressure is reasonably high.

The pressure proportioner uses a pressure vessel filled with high-expansion foam liquid. As water under pressure flows through the proportioner's operating heat, part of it is diverted to the tank to pressurize the foam liquid inside. The liquid is forced through the syphon tube into the low-pressure area of the venturi where it joins the water flowing to the generators. Ideal if water pressure is limited.

The proportioning pump is connected to discharge directly into the water supply. The foam liquid is drawn from the tank and pumped through an orifice plate. The foam solution then passes to the foam generator. This method is often used where large quantities of solution are required. With this proportioner, there is little pressure drop in the system.

Figure 13.2.

density. Occupancy consists of vertically stacked, unbanded rolled kraft paper 25 ft high.

Assume: Fire will open 50 sprinkler heads. Foam leakage around closed doors, drains, etc.; hence, C_L = 1.2

Calculation (References in parentheses are to paragraphs of NFPA-11A):

Foam depth: Depth = 25 × 1.1 = 27.5 ft (this is greater than minimum cover of 2 ft)

Submergence volume: V = 100 × 200 × 27.5 = 550,000 cu ft

Submergence time: T = 5 min (from table)

Rate of foam breakdown by sprinklers:

s = 10 cfm/gpm (from Section 2,352)

Q = No. of heads × area/head × density = 50 × (10 × 10) × 0.25 = 1,250 gpm

R_s = s x Q = 10 × 1,250 = 12,500 cfm

Normal foam shrinkage: C_N = 1.15 (from Section 2,353)

Leakage: C_L = 1.2 (assumption)

Total generator capacity:

$$R = \left(\frac{V}{T} + R_s \right) C_N C_L$$

$$R = \left(\frac{550,000}{5} + 12,500 \right) \times 1.15 \times 1.2$$

R = 169,000 cfm

The number of generators required will depend on the capacity of the generators available.

In calculating the size of the hazard space, permanent nonflammable structures occupying the space are deducted. Space occupied by flammable stock is *not* deducted.

The number of required foam generators for a given system is calculated by dividing R by the capacity (in cfm) of the generator available. In specifying the foam generator, however, water pressure must be considered. Obviously, the total number of generators required will depend on total requirement related to the capacity of individual generators at available water pressure. As a general rule, foam generator capacity increases with increasing water pressure.

References

NFPA 11: *Standard for Foam Extinguishing Systems, 1969.*

NFPA 11A-TR: *Tentative Standard for High Expansion Systems, 1969.*

NFPA 16: *Foam Water Sprinkler Systems and Foam Water Spray Systems, 1968.*

Cray, E. W., "High Expansion Foam," *Fire Journal* (July, 1964).

Lindeken, C. L. And Taylor, R. D., "High Expansion Foam Fire Control Systems for Glove Boxes," *Fire Technology* (August, 1965).

Beers, R. J., "High Expansion Foam Fire Control for Records Storage," *Fire Technology* (May, 1966).

Russell, Roger M. L., "Space Age Protection for High Rise Storage," *The Sentinel* (March, 1967).

Rasbash, D. J. and Langford, B., "The Use of Nets as Barriers for Retaining High Expansion Foam," *Fire Technology* (November, 1966).

Williams, John R., "Effects of Combustion Products on High Expansion Foam," *Fire Journal* (November, 1968).

Jamison, Will B., "Stability: The Key to Effective High Expansion Foam," *Fire Journal* (November, 1969).

National Fire Protection Handbook, 13th Edition (Boston, 1969).

Part III

Special Agent Systems, Extinguishers, AND HVAC Systems

Carbon Dioxide Systems

WALTER M. HAESSLER, P.E.

Most flammable solids, liquids and gases consist principally of carbon and hydrogen, with minor proportions of oxygen, nitrogen, sulfur, etc. Generally, they are defined as hydrocarbons, alcohols, ketones, ethers, esters, carbohydrates, etc., and are most often met within industrial processes.

Flames indicate spatial combustion of gases and vapors distilled from liquids and solids through the action of air diffusion owing to thermal convection currents and the thermal radiation of the flames. Non-flaming combustion (i.e., "glowing") occurs only with solid fuels such as carbon and high-order hydrocarbons that are the residue products after most of the volatile constituents have been distilled away. Although flaming combustion is basically a chain reaction, surface combustion of the deep-embered "glowing" type is not. Extinguishing mechanisms, to be acceptable, must be effective in both instances.

The temperature levels obtained within the outer fringes of the flames are surprisingly uniform, being $3,200°$ to $3,600°$ F. Radiative effects are hence generally alike on a unit basis, and from this situation, we can define a fire as a time rate of fuel combustion expressed as so many Btu's/sec.

Carbon dioxide, if properly applied, is an excellent way to extinguish flames. It is particularly desirable because it is clean compared with wet foam or dusty dry chemical inundations and because it is of low toxic order.

Carbon dioxide exerts a retarding chemical effect on fire, in addition to lowering the concentration of oxygen, by reacting endothermically with free carbon evolved in cracking immediately after combustion starts. Before carbon dioxide is applied, the relatively-slow-burning carbon is seen as black smoke. After the carbon dioxide is applied, the evolution of black smoke is diminished and ultimately eliminated during which time the orange-red flames become blue, the color of burning carbon monoxide.

The concentrations of H* and OH* radicals determine flame velocity and hence the H/C ratio of a fuel becomes a matter of interest. For methane, this ratio is 0.33 and flame velocity is much higher than it is for gasoline, which has a value of 0.19 and much more than for lubricating oils, where the value approximates 0.16. Flame velocity is important when discussing so-called local application means of introducing carbon dioxide into a flaming zone.

In fire extinguishing systems, carbon dioxide is frequently stored in so-called high-pressure cylinders (DOT 3A or 3AA 1,800 or higher) at a pressure of 850 psia at $70°$ F. A maximum temperature of $120°$ F. (equivalent to 2,000 psia), and a minimum of $32°$ F. (equivalent to 490 psia) are set. For ambient temperatures either above $120°$ F. or below $32°$ F, special steps are taken, such as underfilling for high temperatures or adding nitrogen to the cylinder charge ("winterization"). Carbon dioxide may also be stored in so-called low-pressure storage insulated tanks kept in a refrigerated state usually at zero F. (equivalent to 300 psia). This facilitates storage of large quantities—two or more tons in a single tank. (Tanks sometimes contain eighty or one hundred tons for extremely large hazards.)

Methods of Extinguishing

Total Flooding

This represents one of two means of extinguishing. Carbon dioxide is injected into an enclosed space and not directed at the fire. The origin of the fire cannot be predetermined, it being known only that it may be initiated anywhere within the enclosure. After the carbon dioxide is released, either manually or automatically, the space is subsequently purged in a total sense, with the carbon dioxide generally discharged at or near the ceiling, but not necessarily restricted there. Simultaneous venting occurs to the outside atmosphere. The vents must be situated as remotely as possible from the carbon dioxide entry nozzles for conservation purposes.

The concentration of carbon dioxide is raised enough to extinguish the flames. At the start, pure air passes through the vent, but the exiting air later becomes vitiated with carbon dioxide. The amount of carbon dioxide needed varies with the size of the space, the fuels involved and the air ventilation.

Figure 14.1 shows the relationship of flammability of a specific fuel, as it is influenced by the percent of combustible vapor and percent of inert gas in air. With no added inert gas, the upper (rich) and lower (lean) flammability limits are immediately recognized. As the inert gas is added, the space between the upper and lower flammability limits decreases, narrowing the range of combustion activity. Finally, the limits merge at the "nose" of the shaded enclosed area. Combustion will then stop.

Figure 14.2 is a specific example for the case of gasoline vapor. A carbon dioxide concentration of 29 percent is the minimum theoretical value needed to preclude gasoline vapor from burning. For comparative purposes, two other inerting media are shown. If nitrogen were used (not very likely), a minimum concentration of 42 percent in air (88 percent N_2, 12 percent O_2) would be needed to preclude combustion. Nitrogen is known not to enter the combustion reactions except in a minor way, and its role in the example is purely as an oxygen dilutent. The lesser amount of carbon dioxide needed is evidence that, in this case, oxygen dilution is not the sole means of extinguishing, though the greater effectiveness of carbon dioxide, plus the fact that it can be stored in a liquified condition at room temperature, makes it much more desirable than nitrogen. However, sometimes nitrogen is used, if available, for a process, to keep an enclosed space in a constant state of oxygen dilution sufficient to prevent combustion. Such spaces, for obvious reasons, are noninhabited and are normally so protected because a constant hazard of explosion exists. Since the molecular weights of air and nitrogen are almost alike (28), diffusion is rapid, and stratification of these mixtures is impossible. So every location in the space has the same concentration, and a full measure of explosion prevention is obtained.

Also shown in Figure 14.2 is the effect of using bromotrifluormethane (CF_3BR), known commercially as Halon 1301. It is the most effective known halogenated hydrocarbon flooding agent from the standpoint of amount needed to preclude combustion within a given space. It will be observed that only

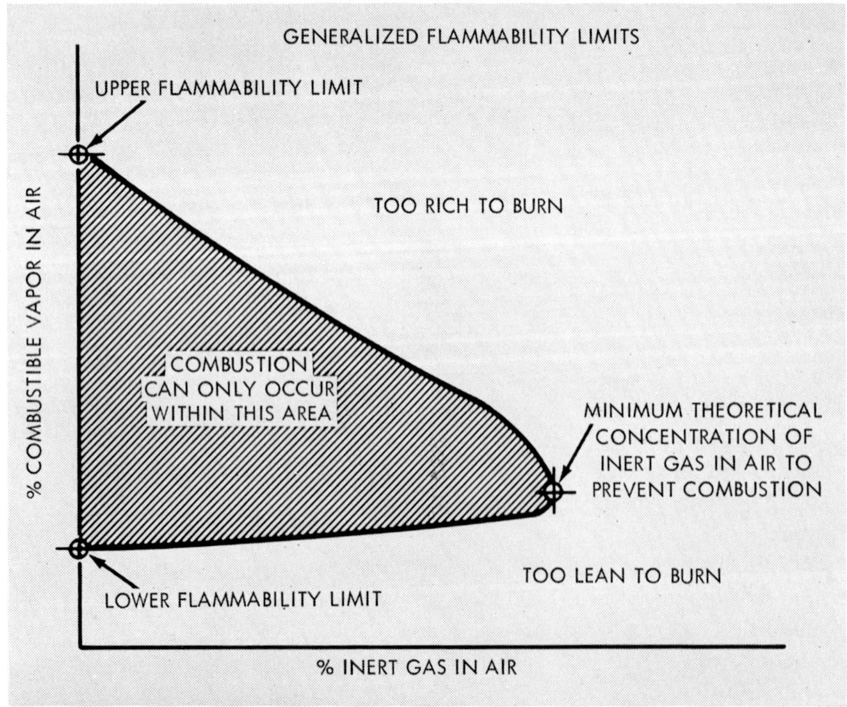

Figure 14.1

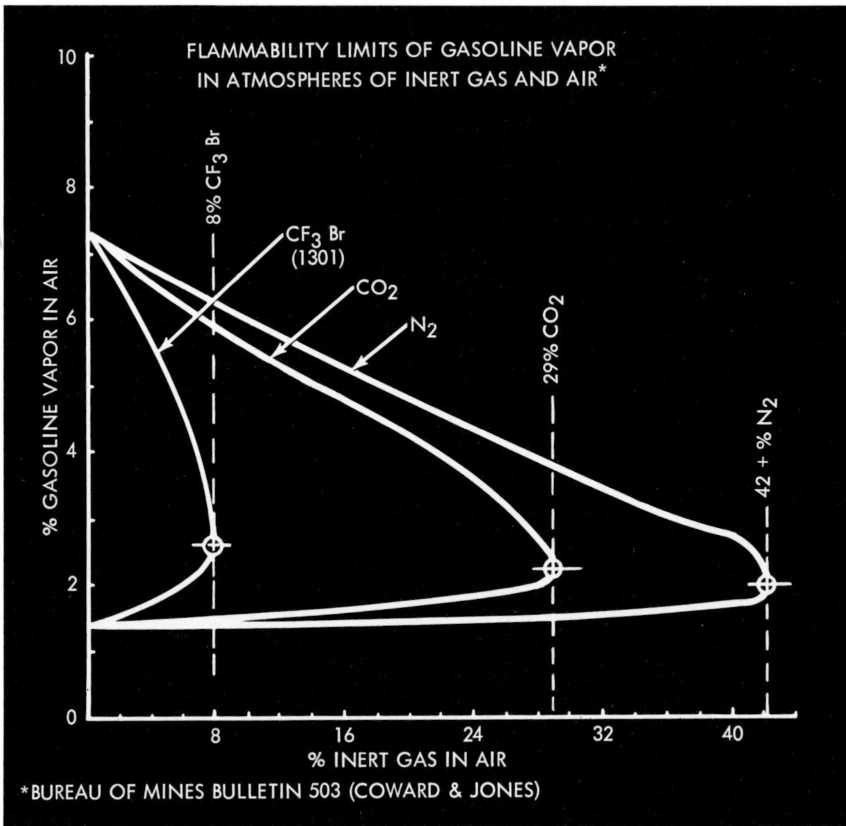

Figure 14.2

8 percent of Halon 1301, on a volume basis in air, will prevent combustion. This is because Halon 1301 reacts with the essential chain reactions during combustion (hydroxylation) removing the hydroxyl (OH*) radicals.

Good practice dictates that the concentration of carbon dioxide will always exceed theoretical concentrations. This is because empirical allowances are needed to compensate for the inevitable lack of perfect diffusion of gases at any moment in any part of the protected space. Table 14.1 defines the theoretical and minimum concentrations of carbon dioxide for several fuels listed in a descending order of flammability.

It has been found that for certain fuels, notably the hydrocarbons of the C_nH_{2n+2} and C_nH_{2n-2} types and of C_2 or higher content, that lesser percentages of carbon dioxide and Halon 1301 extinguish fires than are needed to inert or preclude the possibility of flame existence. That strange condition does not exist in hydrogen, carbon monoxide, alcohols, ketones, carbon disulfide. The situation can point only to still unexplained phenomena in the combustion processes of hydrocarbons, which, with the exception of methane (CH_4), are easier to extinguish than to inert. This condition is practically compensated for by the awareness that from thermal considerations, fires are generally

Table 14.1. Carbon Dioxide Concentrations (% by Volume)

	Theoretical	Minimum Design
Hydrogen	62	74
Acetylene	55	66
Carbon Disulfide	55	66
Carbon Monoxide	53	64
Ethylene Oxide	44	53
Ethylene	41	49
Ethyl Ether	38	46
Ethyl Alcohol	36	43
Butadiene	34	41
Ethane	33	40
Benzol	31	37
Cyclopropane	31	37
Propane	30	36
Propylene	30	36
Isobutane	30	36
Hexane	29	35
Pentane	29	35
Hot Quench and lube oils	28	34
Kerosene	28	34
Gasoline	28	34
n-Butane	28	34
Methyl Alcohol	26	31
Methane	25	30
Ethylene Dichloride	21	25

(Ref: NFPA Pamphlet No. 12; Bureau of Mines Bulletin No. 503; Coward & Jones. The data is based on the minimum concentration of carbon dioxide that will guarantee no flame initiation regardless of the richness or leanness of the fuel-air ratio. This is then the minimum limit of the inerting capability of carbon dioxide.)

harder to extinguish than to prevent. In any event, data shown in the table provide safe and logical bases, backed by experience, for arriving at required inert gas concentrations.

Figure 14.5 shows generalized mathematical relationships of inert gas flow rate, attendant air flow rate due to either natural or forced ventilation, volume of space, elapsed time and resulting inert gas concentrations. It assumes perfect gaseous diffusion wherein the inert gas concentration is the same, at any time, throughout the space. Gaseous diffusion must be quickly achieved by using high-velocity discharge horns positioned to accomplish as uniform a gas concentration as possible through air entrainment. In a fire, thermal convection circulation aids the process in addition to providing more carbon dioxide from the combustion reactions. Practical engineering calculations of the type described are always based on the following:

Concentration of carbon dioxide is achieved by entry from an external supply.

No allowance is made for a reduction in volume, to be protected, owing to any fixed or movable object therein.

Usually carbon dioxide discharge is not to exceed sixty seconds for the usual liquid or gaseous flammable hazards. Regulations sometimes call for extended discharge durations, usually involving rotating machinery that has to be slowed down following shutdown. In certain circumstances, more rapid discharge rates are required to compensate for high ventilating rates that cannot be practically shut down quickly enough or where highly volatile fuels or gases are present.

Normally air ventilation is automatically turned off during discharge to conserve gas and prevent air from being fed to the flames, thereby achieving rapid extinguishment.

Fire extinguishment by total flooding is essentially static. The velocity of inert gas entry is instrumental in achieving rapid diffusion with the air within the enclosure; the confinement of the boundary walls permitted calculations to have meaning; the only important velocity is that of the combustion reactions.

Local Application

Another means of extinguishing, with no confinement in volume, is to blow the flames out, as shown in Figure 14.3. Just as one's breath can blow out a match flame, so can a directed discharge of carbon dioxide do similarly.

A person's breath contains from 3 to 4 percent carbon dixoide and a certain velocity would be needed

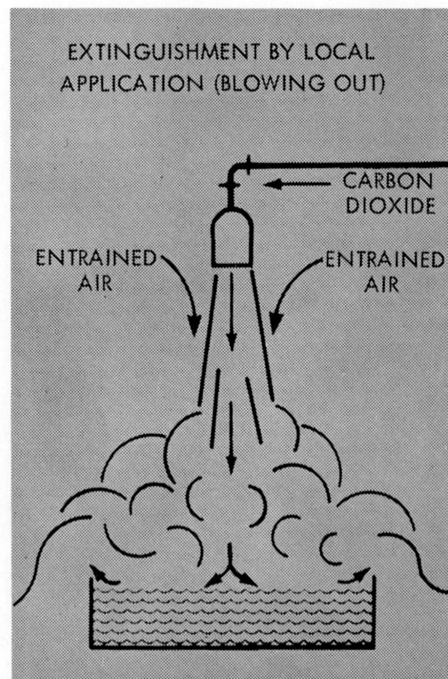

Figure 14.3.

to blow out a flame. If an air jet were used instead, a higher velocity would be needed. If a pure carbon dioxide jet were used, a much lower velocity would be needed. If a pure Halon 1301 jet were used, an even lower velocity would be needed.

For every combination of reactants in a combustion reaction, there is a particular kinetic behavior, which can be referred to as either flame velocity or velocity of flame propagation, and this value is at a maximum level when the reactants are in an exact combining weight proportion (stoichimetric proportions) and when no inert gas is added to the air.

Obviously, the upper and lower flammability limits of combustion must be loci of zero velocity of flame propagation. Figure 14.4 is a three-dimensional extension of Figure 14.2 by adding a "Z" component of velocity. Combustion can only occur within the tetrahedonally bounded enclosure, and extinguishing can be achieved by combinations of velocity, inert gas and flammable vapor outside the enclosure.

Available basic data are too sparse to attack the problem from this direction. However, a wealth of empirical data does exist relating given types of discharge horns, carbon dioxide flow rates, heights of horns above liquid flammable levels and area of flammable surfaces (expressed as square areas for n-heptane preburning for thirty seconds) inside a relatively large, calm enclosure.

The maximum efficiency of extinguishing, expressed as the amount of carbon dioxide needed, occurs at the point of physical incipient fuel splashing, under fire conditions, due to impingement. Normal test criteria limit the carbon dioxide flow rate to 90 percent of the value that would create splashing at a given height

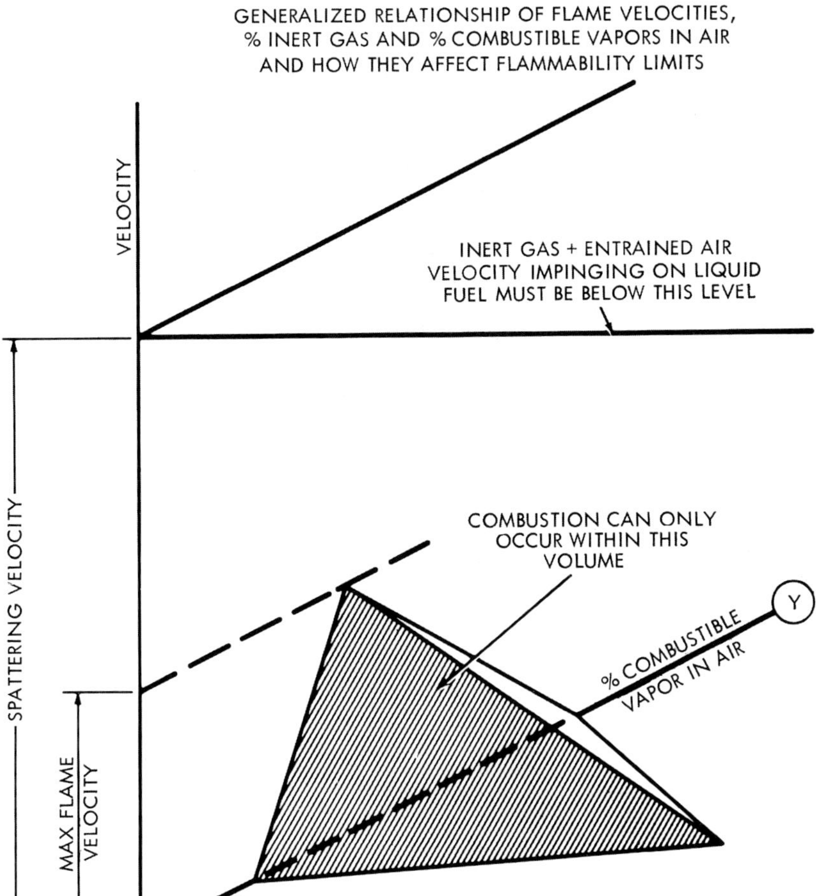

Figure 14.4.

for carbon dioxide cylinder temperatures at 120° F., the fire area capability being determined by what the horn at this height would accomplish if the carbon dioxide cylinder temperature were 32° F. instead.

After initial impingement, the mixture of carbon dioxide and entrained air are directed horizontally and flow radially outward from "ground zero" with an ever-decreasing velocity ultimately reaching the limits of the flammable liquid surface, at which points the velocity of flame propagation must still be exceeded by the dynamic flow-velocity of the carbon dioxide-air mixture. Hence, extinguishing can be achieved over a broad spectrum of the percent inert gas and percent of flammable vapor and velocity, the latter being the reason this process is called dynamic.

Low-velocity horns can be brought close to the fuel surface and produce higher concentrations of carbon dioxide than would be the case for high-velocity horns that would be farther away from the fuel surface and that would produce lower concentrations of carbon dioxide. Extinguishing is thus a dynamic physio-chemical process and an intelligently designed system will result in the fire being blown out with the least amount of carbon dioxide needed.

The extinguishing characteristics of local application nozzles are given in the *Fire Protection Equipment List* of Underwriters Laboratories, Inc. Various manufacturers and factors of flow rate, horn height, and area extinguishing capability are tabulated in the list.

The applications of carbon dioxide fire-extinguishing systems are widespread. They are used to protect generators, transformers, switchgear, dip tanks, drainboards, ovens, quench tanks, rolling mills, solvent storage, oil rooms, paint lockers, printing presses, cargo storage, kitchen ranges, hoods, ducts and dust separators.

Design Considerations

The practical design of a fixed carbon dioxide system requires a working acquaintanceship with basic standards

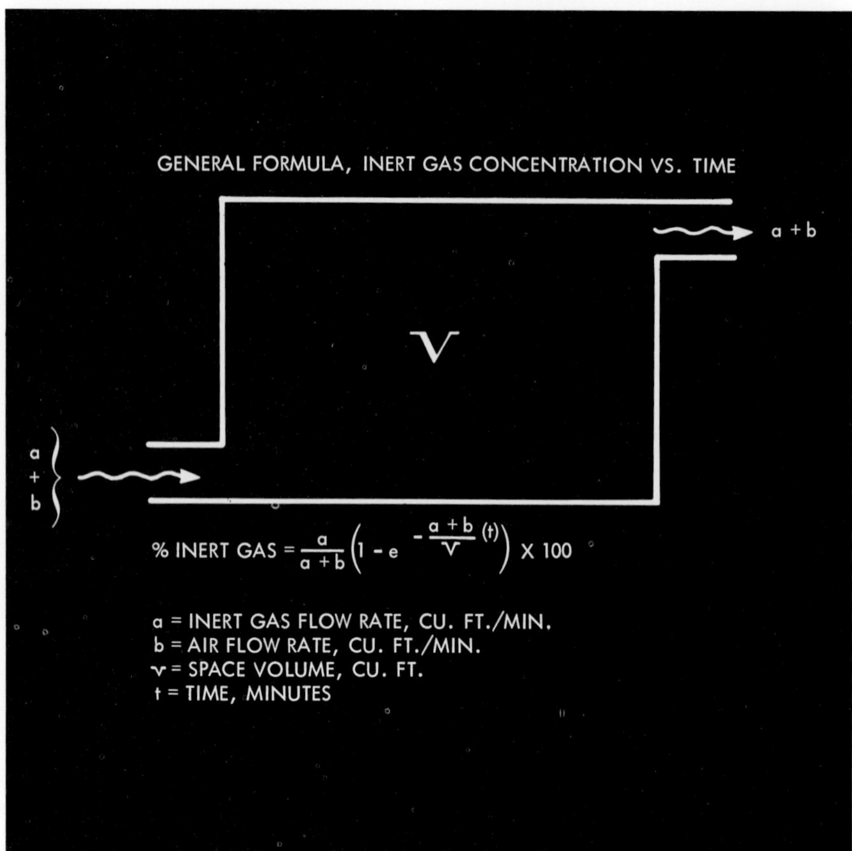

Figure 14.5.

Basic Combustion Chemistry

The mechanism of combustion theory, even for the simplest hydrocarbon, methane, (CH_4), is intricate. If a pool of gasoline is ignited, the vapor above the pool will be the only portion that is burning. This vapor is replenished, as it is consumed, by feedback flame radiative effects exerted on the volatile liquid gasoline. This action brings the fuel in a suitable vaporous form to sustain the flames. (Liquids never burn directly.)

The intense flame radiation "cracks" the distilled vapor into a variety of molecular fragments (radicals), such as Ch_3^+, CH_2^{++}, H^+, C, etc. The fragmentation takes place within the hollow region bounded between the flames and the liquid gasoline pool. The radiative nature of the characteristic orange-red flames is due largely to the incandescent free carbon particles formed in the cracking. This action is easily and readily observed with a spectrometer, which reveals the strong presence of the characteristic carbon lines.

The flaming occurs within a burning zone of a noticeable depth where the entrained air is diffused with the cracked vapors and where the mass ratio of the air with respect to the vapors lies within the lean and rich flammability limits. Within this burning zone, three basic reactions occur:

- Free carbon is believed to turn in four steps:

$$C + \tfrac{1}{2} O_2 \rightarrow CO \text{ exothermic } -47{,}358 \text{ Btu/lb}$$
$$C + \tfrac{1}{2} O_2 \rightarrow CO_2 \text{ exothermic } -122{,}328 \text{ Btu/lb}$$
$$CO_2 + C \rightarrow 2CO \text{ endothermic } +74{,}970 \text{ Btu/lb}$$
$$2CO + O_2 \rightarrow 2CO_2 \text{ exothermic } -244{,}656 \text{ Btu/lb}$$

Total $2C + 2 O_2 \rightarrow 2CO_2$ exothermic $-339{,}372$ Btu/lb

The third reaction is strongly endothermic, absorbing considerably more energy/unit of carbon than is released by the same amount of carbon being oxidized primarily to carbon monoxide, as shown in the first reaction. Carbon dioxide hence can be regarded as an oxidizing agent; or, conversely, the carbon acts as a reducing agent and itself becomes oxidized. This combustion phase is easily

(box copy continued on next page)

discernible. Carbon burns slowly, emits a strong orange-red light and is the main source of infrared flame radiation. Carbon monoxide burns with a blue flame. When carbon dioxide is delivered into flames, the orange-red flame changes to blue, characteristic of carbon monoxide.

• Free hydrogen, the burning of which is better understood, becomes oxidized by branched chain reactions as follows:

$$
\begin{aligned}
H_2 + \lambda &\rightarrow .2H^* \\
H^* + O_2 &\rightarrow OH^* + O^* \\
O^* + H_2 &\rightarrow OH^* + H^* \\
OH^* + H_2 &\rightarrow H_2O + H^* \\
\hline
\text{Total } 2H_2 + O_2 &\rightarrow 2H_2O
\end{aligned}
$$

These reactions occur extremely rapidly and are not influenced directly by carbon dioxide. To break the chain, the extinguishant must react with either OH^*, O^*, or H^*, an impossibility with carbon dioxide.

• Hydrocarbon radicals appear to combine with oxygen in a series of successive stages arriving at final products. The intermediate products each represent an escalation step wherein the carbon-hydrogen bonds are being replaced with carbon-oxygen and hydrogen-oxygen bonds through a chain reaction known as "hydroxylation." This is because of the characteristic hydroxyl radical, which is both formed and consumed in the intermediate steps and which

hence can be defined as a "chain carrier," as follows:

$$
\begin{aligned}
CH_4 &\rightarrow CH_2{}^* + H_2 &&\text{Flame initiation} \\
CH_2{}^* + O_2 &\rightarrow CHO^* + OH^* \\
CH_4 + OH^* &\rightarrow CH_3{}^* + H_2O \\
&&&\text{Flame propagation} \\
CH_3{}^* + O_2 &\rightarrow CH_2O + OH^* \\
CH_3{}^* + O_2 &\rightarrow CH^* + 20H^* \\
&&&\text{Chain branching} \\
CH_2{}^* + OH^* &\rightarrow CH^* + H_2O \\
&&&\text{Flame continuation} \\
CH_3{}^* + OH^* &\rightarrow CH_2{}^* + H_2O \\
CH_2O + OH^* &\rightarrow CHO^* + H_2O \\
CHO^* + O &\rightarrow CO + OH^*
\end{aligned}
$$

The concentration of free radicals of the H^* and OH^* variety determine flame velocity. The volatile hydrocarbons of the long-chain paraffin type are always more easily cracked and hence will always have higher concentrations of the radicals. This is why motor fuels are usually a blend of aromatic fuels that crack less easily together with the paraffin type, with or without the additional slowing down effect of tetraethyl lead. Carbon dioxide cannot directly affect the hydrocarbon chain branching. Dry chemical and bromotrifluoromethane, however, have a strong effect in removing the hydroxyl (OH^*) radical, arresting the combustion reaction during the intermediate steps. This is why the latter agents are used when the occasion demands.

---

and guides and associated reference material, namely:

- NFPA Standard No. 12, *Carbon Dioxide Extinguishing Systems.*
- Factory Insurance Assn., *Interpretive Guide for Carbon Dioxide Systems.*
- Factory Mutual Engineering Assn., *Loss Prevention Data—Special—Protection Systems (4-3).*
- Bureau of Mines *Bulletin #503* (Coward & Jones).

Total Flooding Hazards

A total flooding system consists of a supply of carbon dioxide piped to nozzles that discharge into an enclosed space. The enclosure must be able to retain the gas for a prescribed time and in the concentration necessary for complete extinguishment. The enclosure may be a room, a building, an oven or duct.

The important factors to be considered in designing a total flooding hazard are the volume to be protected and the nature of the fuel which may burn. These determine the quantity of carbon dioxide and application time.

The volume of the space must first be calculated. In calculating the volume, only the volume of permanent nonremovable impermeable structures can be subtracted. The volume taken up by machinery or the ordinary contents of the room cannot be subtracted. The volume when divided by the proper factor obtained from Table 14.2, is used to compute the basic gas requirement.

Flammable Liquid Fires

Flammable liquid fires usually involve surface burning. The quantity of carbon dioxide needed for flammable liquid hazards is determined by two methods.

For flammable liquid fuel fires which can be extinguished by a 34 percent carbon-dioxide design

Table 14.2. NFPA No. 12, Table 5. Volume Factors

(A) Volume of Space (cu ft incl)	(B) Volume Factor		(C) Calculated Quan. (lb) Not Less Than
	(cu ft/ lb CO_2)	(lb CO_2/ cu ft)	
Up to 140	14	.072	—
141 – 500	15	.067	10
501 – 1600	16	.063	35
1601 – 4500	18	.056	100
4501 – 50000	20	.050	250
Over 50000	22	.046	2500

Table 14.3. NFPA No. 12, Table 4. Minimum Carbon Dioxide Concentrations for Extinguishment

Material	Theoretical Min. CO_2 Concentration (%)	Minimum Design CO_2 Concentration (%)
Acetylene	55	66
Acetone	26*	31
Benzol, Benzene	31	37
Butadiene	34	41
Butane	28	34
Carbon Disulphide	55	66
Carbon Monoxide	53	64
Coal or Natural Gas	31*	37
Cyclopropane	31	37
Dowtherm	38*	46
Ethane	33	40
Ethyl Ether	38*	46
Ethyl Alcohol	36	43
Ethylene	41	49
Ethylene Dichloride	21	25
Ethylene Oxide	44	53
Gasoline	28	34
Hexane	29	35
Hydrogen	62	74
Isobutane	30*	36
Kerosene	28	34
Methane	25	30
Methyl Alcohol	26	31
Pentane	29	35
Propane	30	36
Propylene	30	36
Quench, Lube Oils	28	34

Note: The theoretical minimum extinguishing concentrations in air for the above materials were obtained from Bureau of Mines, Bulletin 503. Those marked * were calculated from accepted residual oxygen values.

concentration, standard flooding factors from Table 14.2 (Table 5 NFPA 12) are used. These may include:

1. Paint or ink mixing and storage rooms
2. Solvent storage rooms (or buildings)
3. Chemical process rooms
4. Dynamometer test rooms
5. Test cells or chambers

Some flammable liquid fires require a design concentration over 34 percent to extinguish. These are listed in Table 14.3 (Table 4 NFPA 12) with their minimum carbon dioxide concentrations. Using the design concentration with the curve in Figure 14.6.a (NFPA 12), a conversion factor (from 1 to 4) is reached, by which the basic carbon dioxide quantity of Table 14.2 is multiplied to arrive at a greater quantity sufficient for the particular material. These include:

6. Hazards previously mentioned but containing materials as noted in Table 14.3.
7. Chemical laboratories (sometimes considered to be a standard flammable hazard—sometimes special—depending on contents); *consult authority having jurisdiction*

Openings

The total area of uncloseable openings in square feet should not exceed 3 percent of the total volume in cubic feet. If this area exceeds 3 percent, the hazard usually cannot be total flooded. If under these circumstances the hazard is total flooded, a concentration test is mandatory to prove the system's effectiveness. In general, if 3 percent is exceeded, the hazard should be considered a local application hazard.

It is necessary to close large openings such as doors, windows or fire dampers automatically upon system discharge using pneumatic releases to trip the closing mechanism. Similarly, it is necessary to shut off exhaust fans, pumps or other motor-driven equipment automatically upon system discharge using pneumatic switches or allow extra carbon dioxide for losses produced by continuing operation of such equipment.

When openings cannot be closed, a "screening gag" in the amount of one pound carbon dioxide per square foot of uncloseable opening must be added to the basic carbon-dioxide quantity.

Deep-Seated Fires

The approach to total flooding hazards involving material subject to deep-seated or smoldering fire (such as bulk paper, furs, electrical insulation and electric generators) is similar to the one for surface-burning materials. But differences should be noted.

For "deep-seated" or "Class A" Hazards, flooding factors Table 14.4 (from Table 6 NFPA 12) are used. For example:

1. Blueprint or record vaults
2. Magnetic computer tape rooms
3. Fur-storage vaults
4. Dust collectors
5. Switchgear units or rooms.

While the design concentration must be achieved in one minute for surface fires, this time may be seven minutes for deep-seated fires, provided that at least a

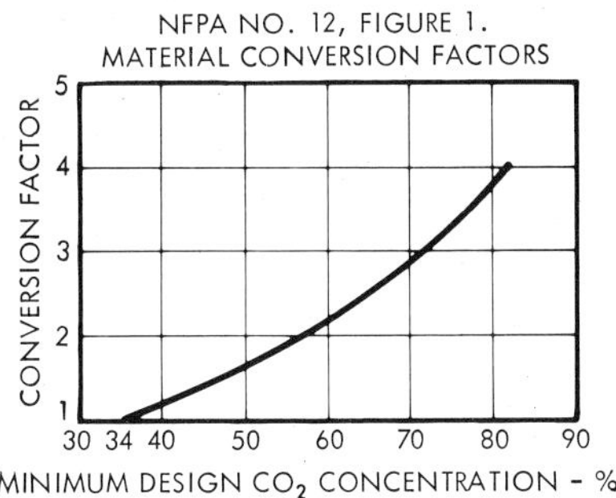

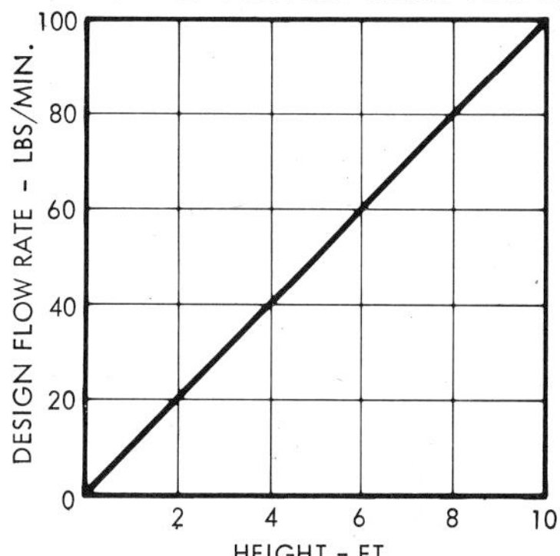

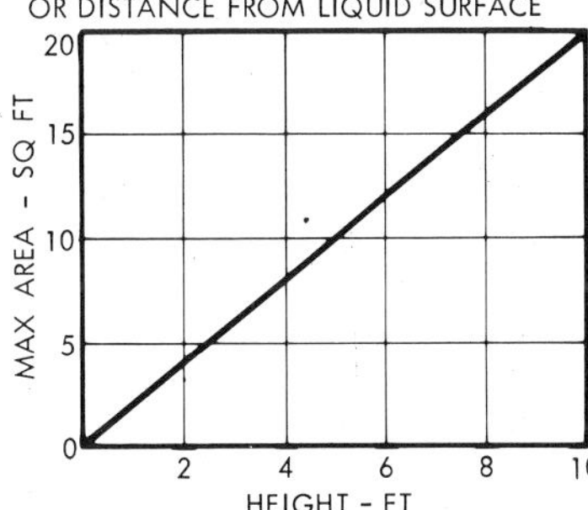

Figure 14.6.

30 percent concentration is achieved in two minutes for a deep-seated fire.

The tightness of the enclosure is more important with deep-seated fires because the carbon dioxide concentration must be greater and must be maintained for a longer time to ensure complete extinguishing. There should be no uncloseable openings except small ones in or near the ceiling. These openings are considered vents and allow air to escape, which averts a buildup of excessive pressure and, if the carbon dioxide is introduced at a level low in the room, results in a higher final carbon dioxide concentration.

Openings through which leakage will occur should be measured carefully so an engineering estimate of leakage can be made. An additional quantity of carbon dioxide would then be determined for an "extended" or "continuous delayed" discharge. An extended discharge is seldom used except with rotating electrical machines.

Whenever a continuous delayed discharge is used, a concentration test should be made to prove the design.

Rotating Electrical Machines

Enclosed rotating electrical machines, such as generators, are protected by an initial carbon dioxide discharge, plus a continuous delayed discharge to allow for deceleration (coastdown) time. The initial gas requirement is determined by dividing the volume of the generator, plus the volume of its connected enclosed recirculating air cooling system, by a factor of ten if the volume is less than 2,000 cubic feet or by a factor of twelve for larger volumes (with a minimum of 200 pounds when the factor twelve is used). The delayed gas requirement is taken from Table A-9 in Appendix A of NFPA No. 12 (see Table 14.5) and depends on the volume protected and the deceleration time of the machine. The deceleration time is what is needed for the machine to coast down (without brakes) from full speed to stop and may be from a few minutes to more than an hour. Usually, the deceleration time is thirty minutes. Generators that have no recirculating air cooling system but are enclosed

Table 14.4. NFPA No. 12, Table 6. Flooding Factors for Specific Hazards

Design Concentration	Flooding Factor (cu ft/lb CO₂)	(lbCO₂/cu ft)	Specific Hazard
50	12	.083	Dry electrical, wiring insulation hazards in general.
50	10	.100	Small elec. machines, wire enclosures, under 2000 cu ft.
65	8	.125	Record (bulk paper) storage, ducts, and mechanically ventillated covered trenches.
75	6	.166	Fur storage vaults, dust collectors.

by means of dampers are so treated, but the delayed gas quantity is increased by 35 percent over the quantity indicated for recirculating the machine.

The volume of a generator and the deceleration time can be obtained from the generator manufacturer. The volume of the cooling system is not usually available from the manufacturer. Obtain the dimensions either by direct measurement or from a drawing of the cooling system, usually available from the hazard owner.

Life Safety

If persons are working in a total flooding hazard, it may be necessary to provide an evacuation alarm before the gas is released into the hazard. In such a case, an alarm and a time-delay device would be required.

Local Application Systems

NFPA 12, Para. A-31 states:

A local application carbon dioxide system is designed to apply carbon dioxide directly to a fire which may occur in an area or space which essentially has no enclosure surrounding it. Such systems must be designed to deliver carbon dioxide to the hazard being protected in a manner which will cover or surround all burning or flaming surfaces with carbon dioxide during operation of the system.

The quantity of carbon dioxide is determined by two methods:

Rate-by-area Method (NFPA No. 12, Section 34) With this method, the quantity of carbon dioxide depends on approval listing of the particular nozzle being used and varies with the distance of the nozzle from the protected surface. In general, the farther the nozzle is from the protected surface, the larger the area covered and the more carbon dioxide required.

Rate-by-volume Method (Section 35) With this method, the quantity of carbon dioxide is based on an assumed enclosure entirely surrounding

Table 14.5.   NFPA No. 12, Table A-9. Extended Discharge Protection for Enclosed Recirculating Rotating Electrical Equipment (Cu Ft Protected for Deceleration Time)

Lbs CO_2	5 Min	10 Min	15 Min	20 Min	30 Min	40 Min	50 Min	60 Min
100	1200	1000	800	600	500	400	300	200
150	1800	1500	1200	1000	750	600	500	400
200	2400	1950	1600	1300	1000	850	650	500
250	3300	2450	2000	1650	1300	1050	800	600
300	4600	3100	2400	2000	1650	1300	1000	700
350	6100	4100	3000	2500	2000	1650	1200	900
400	7700	5400	3800	3150	2500	2000	1600	1200
450	9250	6800	4900	4000	3100	2600	2100	1600
500	10800	8100	6100	5000	3900	3300	2800	2200
550	12300	9500	7400	6100	4900	4200	3600	3100
600	13900	10900	8600	7200	6000	5200	4500	3900
650	15400	12300	9850	8300	7050	6200	5500	4800
700	16900	13600	11100	9400	8100	7200	6400	5600
750	18500	15000	12350	10500	9150	8200	7300	6500
800	20000	16400	13600	11600	10200	9200	8200	7300
850	21500	17750	14850	12700	11300	10200	9100	8100
900	23000	19100	16100	13800	12350	11200	10050	9000
950	24600	20500	17350	14900	13400	12200	11000	9800
1000	26100	21900	18600	16000	14500	13200	11900	10700
1050	27600	23300	19900	17100	15600	14200	12850	11500
1100	29100	24600	21050	18200	16600	15200	13750	12400
1150	30600	26000	22300	19300	17700	16200	14700	13200
1200	32200	27300	23550	20400	18800	17200	15600	14100
1250	33700	28700	24800	21500	19850	18200	16500	14900
1300	35300	30100	26050	22650	20900	19200	17450	15800
1350	36800	31400	27300	23750	22000	20200	18400	16650
1400	38400	32800	28550	24900	23100	21200	19350	17500
1450	39900	34200	29800	26000	24200	22200	20300	18350
1500	41400	35600	31050	27100	25250	23200	21200	19200

the hazard resulting in an imaginary volume to which is applied a rate factor. This factor varies if actual walls are a part of the enclosures.

Typical Local Application Hazards

Area Method–Dip tanks and drainboards, quench and other open tanks, and coating machines.

Volume Method–Printing presses, coating machines and various objects of irregular size and shape that cannot be readily reduced to equivalent surface areas for treatment by the area method. This method is also used if nozzles cannot be positioned as would be required by the area method.

It is often necessary to calculate the gas requirement by both the area and volume methods to determine which results in a lower gas requirement.

Combination of Area and Volume Methods–Dip tanks and drainboards with coated stock suspended over their surfaces in such a manner that the stock is more than two feet above the surfaces. (Stock within two feet of the specified protected surfaces is considered to be protected by the same gas protecting the surfaces.)

The NFPA Code spells out the design factors as follows:

The important factors to be considered in the design of a local application system are the rate of flow, the height and area limitations of the nozzles used, the amount of carbon dioxide needed, and the piping system. The steps necessary to lay out a system are as follows:

1. Determine the area of the hazard to be protected. In determining this area, it is important to lay out, to scale, the actual hazard showing all dimensions and limitations as to placement of nozzles. The limits of the hazard should be carefully defined to include all combustibles which may be included in the hazard, and the possibility of stock, or other obstructions which may be in or near the hazard should be carefully considered.

2. For overhead type nozzles, based on the height limitations of the hazard to be protected, lay out the nozzles to cover the hazard by using various nozzles within the height and area limitations which are expressed in the listings or approvals of these nozzles. The limits on area coverage of a nozzle for a particular height will be determined from listing information which is presented in a form similar

to that shown by Figure 14.6.c [NFPA No. 12, Figure A-9]. In considering the area which is covered by a particular nozzle, it is important to remember that all nozzle coverage is laid out on the basis of approximate squares. Omit this step for tankside or linear type nozzles.

3. Based on the height above the hazard of each nozzle, determine the optimum flow rate at which each nozzle should discharge to extinguish the hazard being protected. This will be determined from a curve such as the one shown by Figure 14.6.b [NFPA No. 12, Figure A-8] which will be given in the individual listings or approvals of nozzles.

4. Determine the discharge time for the hazard. This will always be a minimum of thirty seconds, but may be longer, depending on such factors as the nature of the material in the hazard and the possibility that some hot spots may require longer cooling.

5. Add up the flow rates of the individual nozzles to determine the total flow rate and multiply this by the duration of discharge to determine the total quantity of carbon dioxide needed to protect the hazard. Multiply this figure by 1.4 (for high pressure systems) to obtain total capacity of storage cylinders.

6. Locate the storage tank or cylinders and lay out the piping connecting the nozzles and storage containers.

Miscellaneous Design Items

In the design of any system, attention must be paid to location of storage cylinders or tanks, piping runs, operating hardware, fire detectors, etc. These will vary somewhat based on needs of the hazard and the particular manufacturer's system used.

Calculation Methods

Having completed the design with respect to nozzle location, carbon-dioxide quantity, etc., the remainder of the system design involves sizing of the piping between the storage cylinders or tank and the nozzles. The mathematics of these calculations are precise and complex. For convenience the manufacturers of carbon dioxide have developed and made available calculation curves which simplify the problem.

A sample problem taken from the *Calculation Book for High Pressure Systems* follows.

How to Compute Proper Pipe Size and Nozzle Orifice—High-Pressure Carbon Dioxide

The nature of the fire hazard determines the number and spotting of the discharge nozzles selected on the basis of approved or listed ratings for each particular nozzle. Local conditions determine the position of the storage cylinders and the configuration of the distribution piping. When these factors have been established, a suitable diagram of the system can be prepared.

A simple system serving a hazard requiring three nozzles discharging at the rates noted is illustrated in Figure 1. The design flow rate of the system is the sum of the flow rates through the nozzles served. Each cylinder provides its proportionate share of the total design flow rate to the cylinder bank manifold.

same pipe size is used throughout. It is convenient to tabulate each section as shown in the Sample Computation to simplify the process of determining equivalent lengths and terminal pressures.

The next step is to estimate pipe sizes for each section on the basis of design flow rate and length of run. This is necessary because the method of computation is based on a determination of terminal pressure rather than pipe size. If the selected pipe size turns out to be too small, the pressure drop will be excessive and the final nozzle pressures will be off the chart or much too low. (Generally speaking, nozzle pressures below 300 psia should not be used.) It is then necessary to select larger pipe sizes and repeat the process to determine the proper nozzle pressures. By the same token, if all nozzle pressures are excessively high it is more economical to select smaller pipe sizes.

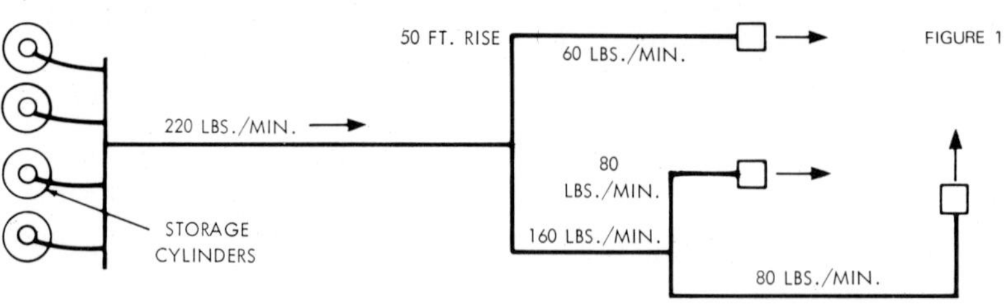

First it is necessary to divide the distribution piping into separate sections having the same pipe size and the same flow rate. For example, the cylinder valve assembly with its flexible connector leading to the cylinder bank manifold is one section. In the illustration, this is labeled A-B. Only one cylinder need be considered since any additional cylinders will be connected in parallel and will have the same flow rate and the same valve and flexible connector assembly. The piping including the manifold to the first branch line will be the next section (B-C) if the

Equivalent Length

For computing purposes it is necessary to use the "equivalent" length of the pipeline rather than the actual length. Pipe fittings, valves or other piping irregularities produce additional pressure drops that can be related to the pressure drop through an equivalent length of straight pipe. The equivalent length of each pipe section is thus the actual length plus the sum of the equivalent lengths of all included bends, pipe fittings and valves.

Equivalent lengths of common fittings in

<table>
<tr><td colspan="6" align="center">SAMPLE COMPUATION</td></tr>
<tr><td>Piping Section</td><td>Pipe Size</td><td>Equivalent Length</td><td>Flow Rate lb/min</td><td>Terminal Pressure psia</td><td>Orifice Size Code No.</td></tr>
<tr><td>A-B</td><td>½ in. Sch. 40</td><td>118</td><td>55</td><td>726</td><td>-</td></tr>
<tr><td>B-C</td><td>1 in. Sch. 80</td><td>235</td><td>220</td><td>630</td><td>-</td></tr>
<tr><td>C-D</td><td>¾ in. Sch. 40</td><td>45</td><td>160</td><td>600</td><td>-</td></tr>
<tr><td>D-E</td><td>½ in. Sch. 40</td><td>20</td><td>80</td><td>570</td><td>7</td></tr>
<tr><td>D-F</td><td>½ in. Sch. 40</td><td>60</td><td>80</td><td>550</td><td>7</td></tr>
<tr><td>C-G</td><td>½ in. Sch. 40</td><td>205*</td><td>60</td><td>550</td><td>6</td></tr>
<tr><td colspan="6">*Section C-G includes a 50-ft rise to new elevation.</td></tr>
</table>

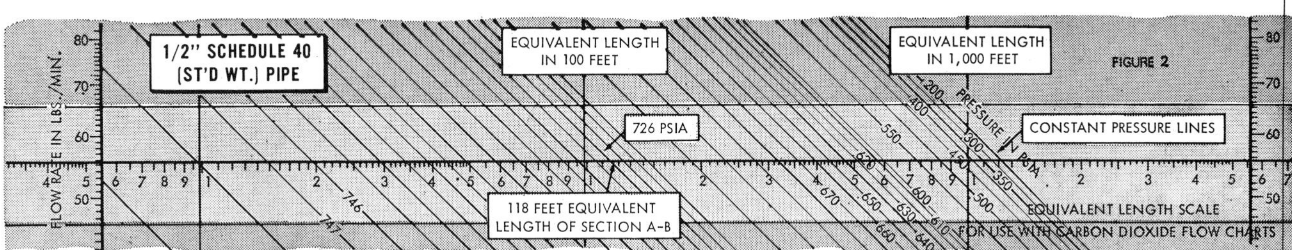

standard sizes are available in published literature such as NFPA 12. The equivalent length of other equipment such as cylinder valves, selector valves, etc., can be found in approvals or listings published by Factory Mutual and Underwriters Laboratories. For the purpose of this example, the "Equivalent Length" given in the sample computation is assumed to include such essential items.

Pressure Drop

To determine the pressure at the end of each pipe section it is always necessary to begin at the storage cylinders where the pressure of the saturated liquid is assumed to be 750 psia. The first pipe section (A-B) in Figure 1 consists of 118 equivalent feet of 1/2 inch schedule 40 pipe (rated equivalent length of cylinder valve and flexible connector). The design flow rate (total rate divided by number of cylinders) is 55 pounds per minute. Referring to the flow chart for one-half inch schedule 40 pipe, the terminal pressure is found by placing the scale at the 55 pounds per minute position and reading the terminal pressure from the inclined pressure lines at the 118 feet equivalent length position. This is found to be 726 psia as illustrated in Figure 2.

The equivalent length scale must be positioned so that the vertical lines on the scale coincide with the vertical equivalent length lines on the chart which are numbered 10, 100 and 1,000. It is also important to keep the scale horizontal by aligning both ends on the desired flow rate.

The pressure at the end of the next section (B-C) is determined in a similar manner except that an additional step is necessary to account for the fact that the starting pressure is 726 psia rather than storage pressure. Pipe section (B-C) consists of 235 equivalent feet of 1-inch schedule 80 pipe and the flow rate is 220 pounds per minute. Referring to the flow chart for one-inch schedule 80 pipe, it will be noted that the 726 psia pressure line crosses the equivalent length scale at 70 feet as illustrated in Figure 3. This then is the starting point to which the 235 feet of section (B-C) must be added. The terminal pressure of this section is thus found to be 630 psia at 305 feet (70 plus 235). By the same procedure the terminal pressure of C-D is found to be 600 psia and the terminal pressures of D-E and D-F are found to be 570 psia and 550 psia, respectively.

Elevation Correction

For nominal changes in elevation in piping the change in head pressure is negligible. However, pipe section C-G includes a rise of 50 feet. The terminal pressure of section C-G is found to be 560 psia by use of the flow chart. The density of the carbon dioxide varies with the pressure, and therefore, the correction per foot of head change depends on the average pressure as shown on Figure 4. Since the average pressure in section C-G is about 595 psia, the correction (from Figure 4) would be about 0.21 pounds per square inch/feet of elevation. The correction is thus 50 X 0.21 or 10.5 pounds per square inch. This must be subtracted from the terminal pressure

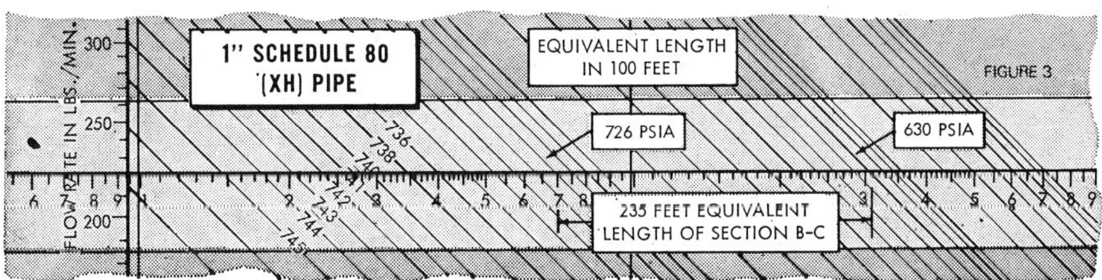

(box copy continued on next page)

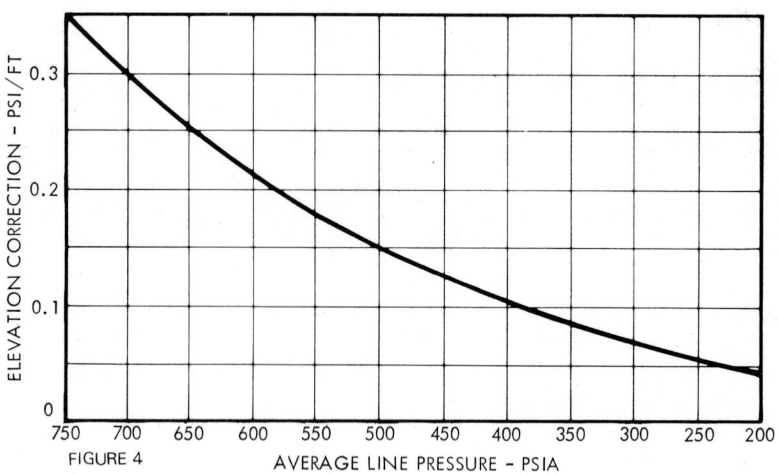

FIGURE 4

determined by the flow chart since there is a loss of head pressure on a rise in elevation. The final terminal pressure of section C-G is thus 560 minus 10 or 550 psia.

Orifice Size

Each discharge nozzle must have the proper orifice size to match the design flow rate at the available terminal pressure. This is determined by use of the orifice flow chart. The same scale used for measuring equivalent length is now used for measuring flow rate in pounds per minute. The scale is placed on the orifice flow chart with the ends aligned on the computed orifice pressure. For example, the pressure at nozzle E is computed to be 570 psia and the flow rate is 80 pounds per minute. From the chart this flow rate at this pressure falls between code number 6 and code number 7 as indicated by the inclined orifice lines. A number 7 orifice would be selected in this case as being closest to the desired flow rate. Similarly, nozzle F also requires a number 7 orifice and nozzle G a number 6 orifice.

Obviously, it is not always possible to select an available orifice size that would exactly match the design flow conditions. The effect of using a slightly larger or slightly smaller orifice is not directly proportional to the flow rates indicated on the chart since any change in flow rate is accompanied by a compensating change in terminal pressure. The flow charts are of course based on equilibrium flow after the pipeline has been cooled to the temperature of the liquid carbon dioxide. The discharge rate of vapor before liquid reaches the nozzles will be considerably less than the final equilibrium or design rate. The delay time before liquid reaches the nozzles and the quantity of liquid vaporized in cooling the pipeline is covered in NFPA 12.

Halon 1301 Systems

CHARLES L. FORD

Halon 1301 fire extinguishing agent has existed for more than twenty years and has been used commercially as a fire extinguishant for nearly ten years. Its safety and effectiveness have stimulated recent emphasis on its use in fire extinguishing systems and portable extinguishers.

The fire extinguishing properties of Halon 1301 were discovered in a series of evaluation tests conducted by the U.S. Army following World War II. First developed for use in a small 2 3/4 pound portable extinguisher for the Army, it was adopted for use in engine fire extinguishing systems on commercial aircraft. The Lockheed Constellation and Douglas DC-7 were the first to use such systems, and every commercial airliner built in the United States since has used the agent. Its weight- and space-saving characteristics have led to its use on private aircraft, battle tanks and hydrofoil watercraft as well. Recent interest has focused on use of the agent in commercial and industrial protection systems.

Halon 1301, a gas under normal conditions, is able to penetrate hard-to-reach places that would form effective baffles for other agents. It is considered a "clean" agent, in that it does not wet or leave a residue on equipment or materials. When applied properly, it is effective on the three main classes of fires—A (common combustibles, such as wood and paper), B (flammable liquids) and C (fires involving live electrical circuits). Since low concentrations of Halon 1301 are required to extinguish most fires, and the agent has a low degree of inhalation toxicity in its natural state, it can be successfuly used to attack fires quickly in normally occupied areas.

Halon 1301 systems range from small half pound units for outboard marine engines to three-and-a-half-ton systems protecting 300,000-cu-ft oil processing buildings. There also are systems for protecting racing car drivers and the crews in armored personnel carriers used by the U.S. Army and Marine Corps. Halon 1301 systems have been used for protecting computer rooms and telephone exchanges, for libraries and museums, for kitchen hoods and ducts, and for many industrial applications. Many small portable extinguishers using Halon 1301 are in service and a few local-application systems have been installed, but most existing systems are of the total flooding type.

The exact mechanism by which Halon 1301 extinguishes fires is not clearly understood, but it is generally thought to be a chemical, rather than a physical, effect (as with water or carbon dioxide). As Haessler [1] and Guise [2] point out, dry chemical and halogenated agents are considerably more effective than strictly heat removal or smothering can account for. The term "chain-breaking" has been applied to the action of these agents.

The combustion of hydrocarbon fuels can be written in four separate chemical reactions. First, the hydrocarbon reacts with oxygen in two steps to form hydrogen peroxide:

$$C_nH_m + O_2 \rightarrow C_nH_m\dot{-}1 + HO_2$$

$$C_nH_{m-1}\cdot + HO_2\cdot \rightarrow C_nH_{m-2}\cdot\cdot + H_2O_2$$

which decomposes into two hydroxyl radicals: $H_2O_2 \rightarrow 2\ OH$. Each hydroxyl radical then can combine with more fuel to form water, one of the final combustion products: $C_nH_m + OH\cdot \rightarrow C_nH_{m-1}\cdot + H_2O$.

This last step is strongly exothermic and accounts for most of the energy produced in a fire. The hydroxyl radicals are considered chain carriers, which propagate the reaction from one fuel molecule to another. If they can be prevented from reacting with fuel, the fire can be extinguished. This results from a lack of sufficient energy being fed back into the first two steps where the fuel and oxygen are prepared for combustion.

When Halon 1301 enters the combustion zone, the following sequence of reactions has been suggested:

First, Halon 1301 is decomposed by heat to produce a bromine radical: $CBrF_3 \rightarrow Br \cdot + CF_3 \cdot$. The bromine radical reacts with fuel to produce hydrogen bromide: $C_nH_m + Br \cdot \rightarrow C_nH_{m-1} + HBr$ which then reacts with a hydroxyl radical to produce water and liberate the bromine radical which can reenter the above reaction: $HBr + OH \cdot \rightarrow H_2O + Br$.

In this process, many active hydroxyl radicals may be removed by a single bromine atom, implying a high degree of effectiveness. As discussed later, this is the case.

Creitz [3] has suggested that bromine atoms present in the combustion zone simply absorb a large number of free electrons by virtue of their large cross sections. Thus, electrons are removed before they can activate oxygen molecules, a prerequisite for reaction. This somewhat different approach illustrates the controversial nature of flame-extinguishment theory.

Regardless of the actual mechanism, only small quantities of Halon 1301 are required to extinguish fires. Table 15.1, from NFPA 12A [4], shows the concentrations of Halon 1301 in air necessary to extinguish and inert many fuels.

Flame Extinguishment vs. Inerting

Two types of flammable liquid or gas hazards must be considered. In one, pure fuel is burning and must be extinguished. In the second, a flammable or explosive mixture of vapors exists and must be prevented from burning. The first effect is called "flame extinguishment," the second, "inerting." Flame extinguishment assumes an established fire of the particular fuel, such as a spill fire or burning gas jet, and the stated concentration is built up surrounding the fire to extinguish it. Halon 1301 concentrations vary from about 2 percent for methane up to 20 percent for hydrogen. Common hydrocarbons and alcohols are extinguished with about 3 percent to 5 percent Halon 1301.

The concept of inerting is somewhat different. Figure 15.1 shows a typical flammability peak curve. With no agent present, the fuel exhibits a lower and an upper flammability limit in air. As extinguishing agent is added, the flammability range of the fuel is

Table 15.1. Recommended Halon 1301 Design Considerations

Material	For Flame Extinguishment	For Inerting
Acetone		5.3
Benzene		4.3
Isobutane	3.6	8.0
N-Butane	2.9	
Carbon disulfide	12.0	12.0
Carbon monoxide	1.0	
Diethyl ether		6.3
Ethane	3.6	
Ethyl alcohol	4.0	4.0
Ethylene	7.2	11.0
N-heptane	3.7	8.0
Hydrogen	20.0	20.0
JP-4		6.6
Kerosine	2.8	
Methane	2.0	2.0
Isopentane		6.3
Petroleum naptha	6.6	
Propane	3.2	6.5
Ethyl acetate		4.6

reduced until at some concentration of agent, it disappears altogether. This concentration of agent is called the "flammability peak" concentration. At higher concentrations of agent, no mixture of fuel and air can be ignited. As might be expected, inerting concentrations are somewhat higher than those for flame extinguishment, ranging from about 4 percent to 8 percent for the common hydrocarbons up to about

TYPICAL FLAMMABILITY-PEAK PRESENTATION

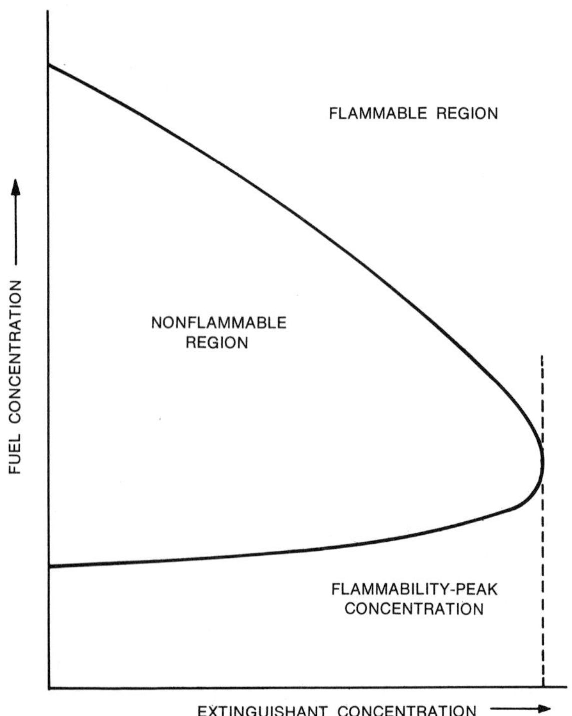

Figure 15.1.

What Is Halon 1301?

Halon 1301* is a chemical compound composed of carbon, fluorine and bromine. Its formula is $CBrF_3$. It is a pure compound and not a mixture of chemicals.

The agent has a high molecular weight, but possesses a higher volatility than is suggested by its high molecular weight. Under normal conditions, it exists as a dense gas, about five times heavier than air. When it is compressed and heat is removed, it can be stored as a liquid having a density about 50 percent greater than that of water. The vapor pressure above the liquid at 70° F. is about 200 psig—considerably lower than for carbon dioxide, but at the same time considerably higher than for other halogenated extinguishing agents. The atmospheric boiling point of the liquid is –72° F.; it will, therefore, vaporize easily in the coldest of climates.

Halon 1301's high volatility does not permit its use in older equipment constructed for carbon tetrachloride or chlorobromomethane. Thus it is impossible to update equipment using these prohibited agents by substituting Halon 1301. The higher vapor pressure of Halon 1301 would probably rupture these lighter containers. In some instances, however, carbon dioxide systems may be converted to use Halon 1301, although the modifications are usually extensive.

*"Halon" is a contraction of the words "halogenated hydrocarbon." In the numbering system originated by the U.S. Army Corps of Engineers to simplify references to these chemicals, the numbers represent, in order, the number of carbon, fluorine, chlorine, bromine and iodine atoms present in the molecule. Terminal zeros (as in the case of Halon 13010) are dropped. Atoms not accounted for are assumed to be hydrogen. The Halon 1301 molecule, therefore, consists of one carbon, three fluorine, no chlorine, one bromine and no iodine atoms: $CBrF_3$.

The physical properties of Halon 1301 are shown below. High dielectric strength, low heat of vaporization, low liquid viscosity and extremely low liquid freezing point further characterize the agent.

Physical Properties of Halon 1301

Molecular weight	148.93
Boiling point at 1 atm. F	–71.95
C	–57.75
Freezing point F	–270.0
C	–168.0
Critical temperature F	152.6
C	67.0
Critical pressure, psia	575.
″ ″ atm.	39.1
Critical volume, cu ft/lb	0.0215
Critical density, lb/cu ft	46.5
″ ″ g/cc	0.745
Specific heat, liquid (heat capacity) at 77° F., Btu/lb-F	0.208
Specific heat, vapor, at constant pressure (1 atm) 77° F. Btu/lb-F	0.112
Heat of vaporization at boiling point, Btu/lb	51.08
Thermal conductivity of liquid at 77° F., Btu/hr-ft F	0.025
Viscosity, liquid at 77° F., centipoise	0.15
Viscosity, vapor, at 77° F., centipoise	0.016
Surface tension at 77° F., dynes/cm	4.0
Refractive index of liquid at 77° F.	1.238
Relative dielectric strength at 1 atm, 77° F. (Nitrogen=1)	1.83
Solubility of Halon 1301 in water at 1 atm, 77° F., wt. %	0.03
Solubility of water in Halon 1301 at 70° F., wt. %	0.0095

20 percent for hydrogen. Most fuels exhibit about a 30 to 40 percent higher concentration for inerting than for flame extinguishment.

NFPA Standard 12A offers the following guidelines for choosing between flame extinguishment and inerting. Flame extinguishing values may be used if:

(a) The quantity of fuel permitted in the enclosure is less than that required to develop a maximum concentration equal to one-half of the lower flammable limit. . . .

(b) The volatility of the fuel before the fire is too low to reach the lower flammable limit in air (maximum ambient air temperature or fuel temperature does not exceed the closed-cup flash point temperature) and the fire may be expected to burn less than 30 seconds before extinguishment.

Unless the above conditions can be met, the inerting level must be applied.

Class A Fires

These are generally considered to involve solid combustibles, mainly cellulosic in nature, such as wood, paper and textiles. For convenience, plastic articles, sheet and foam have also been included in Class A, though these materials have different burning characteristics than cellulosic materials. Most plastics do not develop the degree of char that cellulosics do. In fact, combustion characteristics vary widely between cellulosic materials. So also do the extinguishing characteristics of fires in these materials when Halon 1301 is applied.

Most cellulosic materials burn in two stages. First, flaming combustion can occur along the surface of the fuel array; second, glowing combustion can occur along the surface of the array, within the bulk of the fuel, or both. Glowing combustion usually results from the material first undergoing flaming combustion, although a small amount of glowing combustion can precede flaming. For at least a portion of the fire's life, glowing and flaming combustion take place simultaneously. Although a few plastic materials have been found to exhibit glowing combustion, most undergo flaming combustion only.

The term "deep-seated" is applied to fires that have reached the glowing stage, particularly where the glowing embers have burrowed within the bulk of the fuel. The tendency to become deep-seated increases with length of exposure to flaming combustion (preburn) for most combustibles. Cellulosic fibers (such as cotton), pressed fiberboard, and loosely packed shredded paper are examples of materials that can rapidly develop deep-seated fires. Stacked paper and pressed cardboard require prolonged exposure to flames before the fire becomes deep-seated. In other materials—charcoal, for example—although glowing combustion occurs readily, it remains at the surface of the fuel.

When Halon 1301 is applied to a Class A fire, the flaming combustion can be extinguished with low concentrations of agent—about 3 percent by volume. Glowing combustion may continue, however, at least for a while. If the glowing occurs only on the surface of the fuel, as in a wood or charcoal fire, and the Halon 1301 atmosphere is maintained around the fire, the fire will eventually go out. This period of time, referred to as the "soaking" time, depends on the Halon 1301 concentration used. Figure 15.2, which appeared in the tentative NFPA 12A Standard (1968), shows the relationship between soaking time and Halon 1301 concentration necessary to extinguish charcoal fires.[5]

If combustion has become deep-seated, low concentrations of Halon 1301 will not completely extinguish the fire. Flaming combustion and glowing combustion on the fuel surface will be extinguished. The glowing combustion that occurs *within* the fuel mass, however, will continue, eventually consuming the fuel from inside out. Apparently not enough Halon 1301 molecules are able to diffuse into the fuel mass to be effective. These fires can be extinguished with higher concentrations of Halon 1301, around 20 to 30 percent by volume, but application of these levels is seldom practical.

The latest NFPA 12A Standard distinguishes between deep-seated and surface fires in a functional way. If, from experimentation, the fuel is not completely extinguished with low concentrations of Halon 1301 (say, 3 to 7 percent) that have been maintained for short soaking times (say, 5 to 10 minutes), then the fuel is considered to exhibit deep-seated burning characteristics. Under these conditions, the authority having jurisdiction must decide whether to permit low concentrations of Halon 1301 to be used to control the fire until outside help can arrive or whether to require that the agent extinguish the fire completely. His judgment must include an estimate of how likely the fire is to become deep-seated, considering the method of detection, size and configuration of the

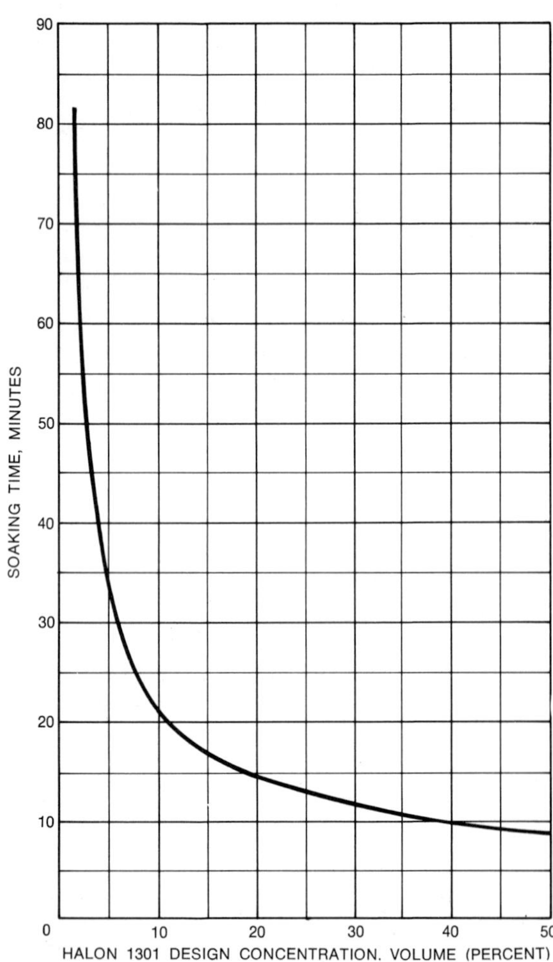

Figure 15.2.

enclosure, ventilation, etc. Requiring complete extinguishment under the worst possible conditions may rule out the use of Halon 1301 in particular applications.

It is apparent that the Class A fire problem is not well understood. Many Halon 1301 equipment manufacturers, as well as the producers of this and other halogenated agents, are continuing research in this area. It seems likely that the treatment of Class A fires in the standard will be modified frequently as more knowledge is acquired.

What Halon 1301 Won't Do

It is important to understand the limitations of Halon 1301. The agent will not extinguish burning metals, metal hydrides or materials that contain their own oxidizing agent (such as cellulose nitrate, gunpowder, solid rocket propellants, etc.). Although the agent will extinguish Class A fires, special design considerations must be observed to make the system effective. High agent concentrations, long soaking times, or both, are required for combustibles likely to become deep-seated. Agent decomposition can sometimes be a problem but can be minimized through proper system design.

Halon 1301 is expensive compared to other common agents; so its use in a system subject to frequent discharge may be uneconomical. (The initial installed cost of the sytem will probably be competitive with systems using other agents, however.)

The Halon 1301 concentrations in air needed to extinguish many fuels may be inhaled safely for short periods of time, but in situations that need high agent concentrations, special safety precautions to preclude personnel exposure are required.

The design of a Halon system must match hardware and agent to a hazard to an even greater degree than for systems using other agents.

NFPA Standard No. 12A (1973) offers detailed engineering data for designing, installing and maintaining Halon 1301 systems. Proper use of the standard requires prior knowledge of fire extinguishing systems, plus an understanding of the properties of Halon 1301 and of the principles on which the standard is based.

Halon 1301 Storage Containers

As attractive as the agent may appear to be in its own right, its performance largely depends on the system in which it is used. Figure 15.3 shows a basic extin-

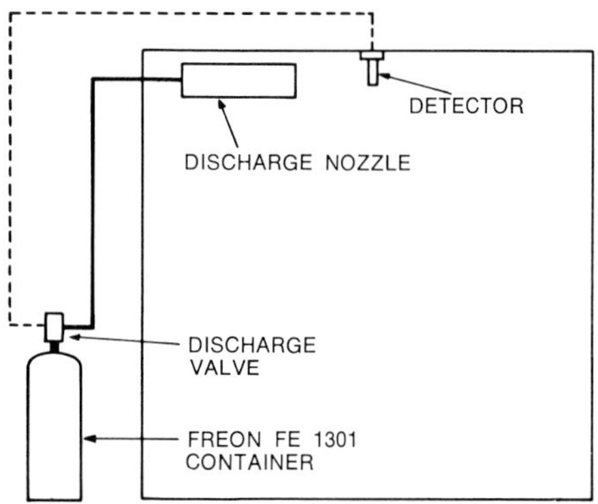

Figure 15.3.

guishing system in schematic form. The system consists of a container of agent arranged to discharge through nozzles to release an adequate quantity of agent to extinguish the fire. A suitable detector may be connected to the release valve to operate the system automatically. The system can be either total flooding, as shown, or local application. Some general requirements and characteristics apply to both system types.

The vapor pressure of Halon 1301 at room temperature is approximately 200 psig. To store the agent in liquified form, a high-pressure vessel, such as a sphere or a carbon dioxide or ammonia type of cylinder is used. When the filled container is shipped interstate, as is the usual practice, the U.S. Dept. of Transportation (DOT) requires that the liquid portion of the contents not completely occupy the container at 130° F. In other words, the container shall not be liquid-full at or below 130° F.[6]

The liquid density of Halon 1301 at room temperature is about 98 pounds per cubic foot. This figure varies with temperature, and at 130° F. it is reduced to about 77 pounds per cubic foot. NFPA 12A, however, requires a 75-pound per cubic foot maximum-fill density (pounds of agent per cubic foot of container internal volume).

The presence of nitrogen creates several undesirable side effects due to its high degree of solubility in liquid Halon 1301. Figure 15.4 shows the solubility constant as a function of temperature. At 70° F., the constant is about 210 psi per weight percent dissolved nitrogen. A 600-psi superpressure level gives about 2 percent nitrogen by weight (600 psi total pressure, minus 200 psi vapor pressure of Halon 1301, divided by 210, the solubility constant) dissolved in the liquid phase. This is a rather high solubility.

One undesirable effect of the presence of the nitrogen is that it decreases the liquid density of the agent. At the 360-psi superpressure level, the effect is small. At 600 psi, however, the liquid density at 130° F. is

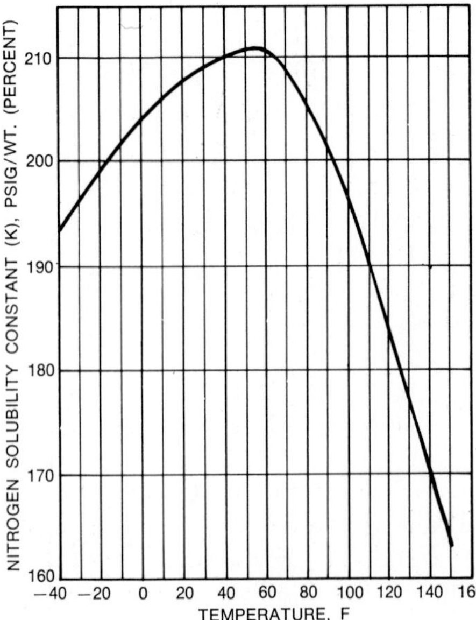

Figure 15.4.

reduced to 72.7 pounds per cubic foot, so that the maximum fill density at this pressurization level must be reduced to 70 pounds per cubic foot to avoid being liquid-full at or below 130° F.

DOT also requires that the pressure of the container at 130° F. does not exceed 5/4 times the service pressure of the cylinder. Many systems, particularly the early ones, were constructed using carbon dioxide cylinders. The pressure rating of carbon dioxide cylinders is high enough (DOT, 3A- or 3AA-1800 or higher) that the pressures generated by Halon 1301 are of no concern. In fact, the 600-psi level was established when Halon 1301 systems were using carbon dioxide equipment exclusively. At lower superpressure levels, less expensive welded cylinders may be used. Since DOT permits a maximum service pressure of 500 psi for 4B or 4BA specification cylinders, the internal pressure at 130° F. can range up to 625 psig (5/4 times 500). Calculating this back to 70° F. and allowing for a 5 percent error in filling results in the 360-psi pressure level. In practice, the fill density is reduced to 68 pounds per cubic foot when using DOT 4B or 4BA-500 containers to avoid overpressurizing at 130° F.

Halon 1301 Flow Characteristics

Upon releasing the agent from the container, the contained pressure above the liquid phase provides the driving force for flow through the piping system and discharge nozzles, where the agent is expelled into or onto the hazard. In engineering fire extinguishing systems, it is important to be able to calculate flow rates accurately. These calculations are difficult in two-phase flow systems, such as a carbon dioxide system.

The combination of nitrogen superpressurization of Halon 1301 and a lower limit on pipeline pressure of 200 psia simplifies the flow calculations by keeping the agent in the liquid phase during its flow through the piping system. In practice, as the pressure exerted on the liquid Halon 1301 is reduced below that present in the storage container, nitrogen separates from the liquid Halon 1301, so that some two-phase flow actually occurs. The nitrogen does seem to prevent gross flashing of liquid Halon 1301 into vapor.

The discharge rate of the system is established by the selection of discharge orifices. If the Halon 1301 flow were single phase during flow through the piping and at the orifice, this calculation would be straightforward. However, the nitrogen dissolution is great enough so that these determinations must be based on experimental data. NFPA 12A includes a table of orifice flow factors, shown in Table 15.2. These factors depend on the initial pressurization level of the container, which establishes the quantity of nitrogen dissolved in the liquid, and on the pipeline pressure immediately above the orifice. This latter value determines the quantity of nitrogen that has separated from the liquid, as well as the driving force for flow through the orifice. The driving force, in turn, depends on the pressure drop that has occurred in agent flow from the storage container to the orifice.

There are three sources of pressure drop in a Halon 1301 system. One is an initial effect of filling the pipeline from the container. This effect can be calculated by considering isothermal expansion of the vapor space in the container to the vapor space that exists after the pipeline has been filled. This step is necessary before liquid Halon 1301 can be discharged from the orifice.

A second source of pressure drop is friction loss due to flow through the system after the piping has been filled. Figure 15.5, from NFPA 12A, shows the pressure drop vs. flow characteristics for Halon 1301 in steel pipe. A similar chart is included in the standard for friction losses in copper tubing. The two-phase effect can be seen from the slight curvature of the lines. Although the effect is small at low values of $\Delta P/\Delta L$, it becomes great enough at high values of $\Delta P/\Delta L$ to preclude the use of classical flow equations of the Fanning or Moody types. Each chart may be used for both levels of nitrogen superpressurization.

The third source of pressure drop is loss due to elevation in piping. This effect is approximately 0.7

Pressure above nozzle, psig	Discharge rate, lb/(sec) (sq in.)*	
	At 600 psig cylinder pressure	At 300 psig cylinder pressure
200	10.3	19.3
210	11.3	21.4
220	12.4	23.5
230	13.5	25.9
240	14.6	28.3
250	15.8	30.8
260	17.1	33.4
270	18.4	36.1
280	19.7	39.0
290	21.1	41.9
300	22.5	45.0
310	24.0	48.1
320	25.5	51.4
330	27.1	54.8
340	28.7	58.2
350	30.4	61.8
360	32.1	65.5
370	33.9	
380	35.7	
390	37.5	
400	39.4	
410	41.4	
420	43.4	
430	45.4	
440	47.5	
450	49.7	
460	51.8	
470	54.1	
480	56.4	
490	58.7	
500	61.1	
510	63.5	
520	65.9	
530	68.5	
540	71.0	
550	73.6	
560	76.3	
570	79.0	
580	81.7	
590	84.5	
600	87.4	

*Nozzle discharge coefficient = 0.98

per square inch/foot increase in elevation. In cases with piping rises of less than twenty feet, the effect is usually ignored. Of course, if the flow is downward, rather than upward, a pressure gain occurs.

Total Flooding Considerations

A total flooding system discharges agent into an enclosure around the hazard to create a uniform concentration of agent throughout the enclosure. This concentration must be adequate to extinguish a fire no matter where the fire is located within the enclo-

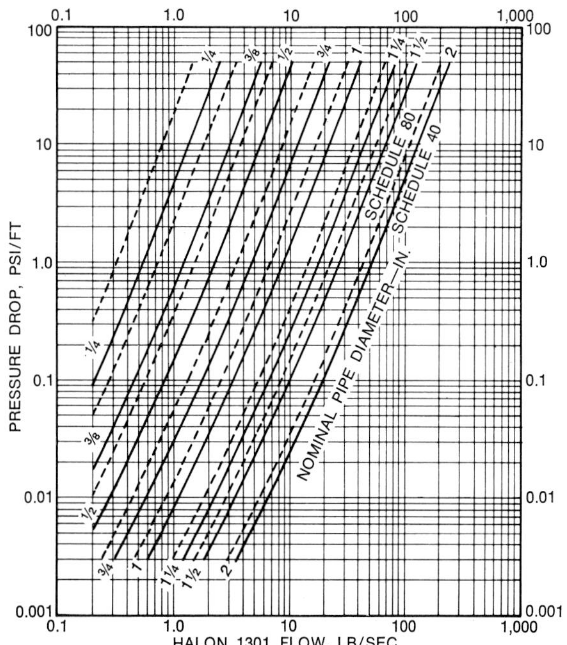

Figure 15.5.

sure. The quantity of agent needed to develop a given concentration is determined by taking into account the volume and temperature of the enclosure, the fuel and type of protection required, and losses through unclosable openings and losses caused by ventilation. (The extinguishing and inerting concentrations required for many typical fuels are shown in Table 15.1.)

Once the design concentration of Halon 1301 has been established, the design quantity can be calculated through an intermediate variable called the "flooding factor." The flooding factor is the number of cubic feet of enclosure volume protected by each pound of agent discharged into it. At room temperature and pressure, the expansion of each pound of Halon 1301 produces 2.56 cubic feet of pure vapor.[7] This value is the specific volume, denoted by v in the following equation for flooding factor:

$$F_v = \frac{V}{C}(100 - C)$$

Where: F_v = volume flooding factor, cubic feet protected volume/pound Halon 1301; v = specific volume of Halon 1301, cubic feet/pound, at temperature and pressure of enclosure; and C = Halon 1301 design concentration, percent by volume.

For design temperatures other than 70° F., the value of specific volume for Halon 1301 is different from the 2.56 cubic feet/pound value given above. It can then be estimated from the following formula, which is reasonably accurate over the temperature range of –65° to +250° F.: v=2.2062 + 0.005046t

Where: t = design temperature, F.

The agent discharge time must be considered in total flooding systems. NFPA 12A requires that the liquid portion of the agent supply be discharged within ten seconds, unless a longer discharge time is specifically permitted by the authority having jurisdiction. The ten second time is intended to minimize decomposition of the agent by reducing the time in which the agent can be exposed to flames.

In situations where the engineer feels that a longer discharge time may be safely used, he may do so provided the authority having jurisdiction agrees with him. Such a situation might be one in which the burning surface would be limited to a small area in proportion to the enclosure volume or where the fire would be detected at an early stage before becoming fully developed. Some situations, conversely, may require a discharge time considerably less than ten seconds— where large quantities of flammable gases may be released, for example. For such areas, suppression systems capable of detecting and extinguishing incipient explosions and of reacting and discharging within a few milliseconds may be required. An element of judgment is, of course, needed in selecting the proper discharge time.

Because Halon 1301 exists normally as a gas, it is subject to the effects of ventilation and leakage through enclosure openings. NFPA 12A gives the formulas necessary to make these corrections. Ventilation of the enclosure is treated much like the calculation of flooding factors. Assuming a constant ventilation rate and perfect and instantaneous mixing of the agent in the enclosure upon discharge, the discharge rate necessary to develop a given concentration within a given time can be calculated by the following formula:

$$R = \frac{CE}{v\left[1 - e^{-\frac{E}{V}t}\right]}$$

Where: R = Halon 1301 discharge rate, pounds per second; E = ventilation exhaust rate, cubic feet per second; V = enclosure volume, cubic feet; t = time, seconds; e = natural logarithm base, 2.71822; C, v as defined above.

If it is necessary to maintain the concentration for a longer period, as for deep-seated fires, an extended discharge of agent is necessary. Otherwise, the agent concentration will begin to diminish as soon as the discharge is terminated.

Because the method of calculating the required quantity of agent differs in ventilated cases, a slight discrepancy occurs at low ventilation rates. The quantity of agent required with ventilation is less than without ventilation. The solution here is to ignore the effects of ventilation if the agent discharge rate (cfm) exceeds the ventilation exhaust rate (also cfm).

Losses through enclosure openings must be treated differently. During the discharge period, losses through openings generally can be ignored since the agent application rate is usually much higher than losses could be through openings, particularly small ones. The period of protection following discharge, however, depends greatly on losses through openings. Since Halon 1301 vapor is about five times heavier than air, a 5 percent concentration in the enclosure has a specific gravity about 25 percent greater than air. The heavier Halon 1301/air mixture in the enclosure is displaced with lighter air from outside the enclosure. In a quiescent state, Halon 1301/air mixture exhausts from the enclosure through the lower half of the opening and fresh air enters the enclosure through the top half of the opening and collects at the top of the enclosure. Factory Mutual Research Corp.[8] has found that a sharp interface is formed between the fresh air and the Halon 1301/air mixture, and that this interface descends with time much as the surface of water descends in a container that has a hole in the bottom. The rate at which the interface descends is a function of the height of the interface above the opening, the size of the opening itself and the concentration of agent originally in the enclosure. An increase in any of these three factors increases the rate of descent. This effect cannot be compensated for by giving the system an initial overdose of agent. In fact, a higher initial concentration results in a more rapid loss of protection in the enclosure.

These losses can be compensated for in three ways. The best method technically is to have an extended discharge of agent equal to the losses through the opening. In addition to maintaining the proper concentration of agent, the agitation caused by the continuing discharge will also maintain a uniform concentration. The extended rate required to maintain a constant concentration is not nearly so great as the initial rate required to achieve it. Second, the concentration can be maintained by mechanically agitating the agent/air mixture in the protected space, usually with a fan. No further agent makeup is needed, but the concentration will gradually diminish at a rate less than without agitation. Third, the interface may be allowed to fall naturally to some predetermined point. This approach should be used only when the hazards are located toward the floor of the enclosure. The details for calculating all three methods of compensation are given in NFPA 12A.

Another important consideration in total flooding systems is the type of discharge nozzle used. Since Halon 1301 vapor is five times heavier than air and the concentrations of agent are generally low, the problem of mixing, particularly in large enclosures, might seem difficult. In practice, however, this is not the case. Many types of nozzles have been used satisfactorily. These include a free jet, carbon dioxide-type baffle nozzles, spray nozzles of various types and air aspirating nozzles. In fact, almost any nozzle that breaks up the liquid stream into a fine pattern will suffice. About the only nozzle found *not* to be satisfactory is the multidischarge nozzle commonly used in carbon dioxide local application and total flooding systems. In all cases, the nozzle should be listed by Underwriters Laboratories or Factory Mutual. If there is any doubt, a discharge test and agent concentration check may be required after the system is installed.

Safety Properties of Halon 1301

The safety of Halon 1301 has been established through more than twenty years of medical research involving both test animals and humans. In discussing its safety, three aspects must be considered: the natural, or undecomposed, agent; the agent's products of thermal decomposition; and the combustion products of the fire itself.

Natural (Undecomposed) Agent
The Army Chemical Center [10] determined the approximate lethal concentration (ALC*) of natural Halon 1301 vapor to rats to be 832,000 ppm (83.2 percent) by volume in air for a fifteen minute exposure. Underwriters Laboratories, Inc., [11] confirmed this low toxicity and placed Halon 1301 in the least-toxic category, Group 6, of its classification of comparative life hazards of chemical vapors. Kettering Laboratory [12] exposed animals to Halon 1301 concentrations, up to 32 percent for seven hours without harm. Paulet [13] performed acute exposures upon rabbits, mice, guinea pigs and rats with Halon 1301 concentrations up to 80 percent (20 percent oxygen) for two hours without mortality, and chronic exposures at 50 percent Halon 1301 (maintaining 20 percent oxygen) for two hours a day for fifteen consecutive days without apparent evidence of toxicity. Hazleton Laboratories [14] exposed animals, including primates, in concentrations up to 20 percent Halon 1301 in air for two hours with no observable changes in blood chemistry during or after exposure.

It is known that certain compounds, such as hydrocarbons, chloroform, etc., can sensitize the heart to high levels of circulating adrenalin in the blood stream. This can result in irregular heart action and, in severe cases, death. In light of this, the sensitizing effects of Halon 1301 to adrenalin on the heart have been studied by several investigators.

Haskell Laboratory [15] produced cardiac arrhythmias in dogs exposed to 7-1/2 percent and higher levels of Halon 1301 for five minutes, when the dogs were given a challenge injection of adrenalin during the last nine xeconds of the exposure. No irregularities were observed up to 20 percent Halon 1301 without the adrenalin injections. Aerospace Medical Research Laboratories [16] demonstrated spontaneous cardiac arrhythmias in both dogs and monkeys at 40 percent or more Halon 1301. Hine Laboratory [17] was unable to produce ventricular fibrillation in dogs at concentrations up to 40 percent Halon 1301 with adrenalin released endogenously by fright, although convulsions and tremors were observed at higher agent concentrations.

Encouraged by the relatively low acute toxicity of Halon 1301 as determined in animal tests, two laboratories have conducted exposures to humans. Haskell Laboratory [18] exposed three persons to Halon 1301 concentrations up to 10 percent for three to four minutes. Hine Laboratory [17] exposed ten subjects to Halon 1301 concentrations up to 17 percent and for times up to twenty minutes. Subjects exposed to higher than 7 percent Halon 1301 experienced feelings of light-headedness and reduced dexterity, which increased in intensity with increasing agent concentration. At concentrations below 10 percent, the effects were not considered to be serious. Exposures to concentrations greater than 10 percent carry a degree of risk to personnel and should be avoided.

NFPA Standard 12A permits Halon 1301 to be used in total flooding systems in normally

*The minimum concentration of a vapor required to produce death in one or more of a group of test animals during or following a specified exposure time.

(box copy continued on next page)

---

occupied areas without restriction, provided the agent design concentration does not exceed 7 percent. Where egress can be accomplished within one minute, design concentrations up to 10 percent may be used. Areas not normally occupied, but where personnel may occasionally be present, can use concentrations up to 15 percent, again provided egress time does not exceed one minute. In situations requiring either higher concentrations or longer egress times, auxiliary breathing apparatus should be provided inside the hazard to protect personnel.

Products of Thermal Decomposition

In extinguishing a fire, part of the Halon 1301 is decomposed from exposure to flames and hot surfaces above about 950° F. The types and concentrations of these by-products varies with fuel type and fire size, enclosure size and extinguishment time. The predominant decomposition products are hydrogen fluoride (HF) and hydrogen bromide (HBr). In past studies, small quantities of free bromine (BR_2) and carbonyl halides (carbonyl fluoride, COF_2, and carbonyl bromide, $COBr_2$) have been observed, although more recent testing has failed to confirm the presence of these materials. Hydrogen fluoride and hydrogen bromide are primarily respiratory, eye and skin irritants, but they have been found from fire tests to be present in concentrations well below their ALC values (estimated below) for fifteen-minute exposures:

Compound	15-min. ALC ppm by vol.	Typical maximum concentration produced in fire tests ppm by volume
HF	2,500	0-500
HBr	4,750	0.50

At these typical levels, both HF and HBr have an irritating effect on personnel. They give the post-extinguishment atmosphere a distinctive sharp, acrid odor. In extreme cases, personnel may find that breathing becomes difficult, and their eyes and skin sting. For exposures of only a few minutes, this atmosphere does not present a severe hazard, aside from possible individual reaction to this discomfort. On the positive side, the odor can serve as a warning to personnel who might otherwise attempt to enter an area where there is a fire.

System design can play a large part in minimizing decomposition of Halon 1301. First, the size of the fire can be kept to a minimum through use of the proper type of detector for the hazard. This is not to say that the most sensitive detector per se should always be used; the most sensitive detectors are often the most prone to false alarm. Rather, some optimum point between sensitivity and reliability should be sought, and this particular point will vary from hazard to hazard. Second, once the system is activated, the agent concentration necessary to extinguish the fire should be established quickly. The shorter the agent discharge time, the less contact between Halon 1301 and flames; consequently, the decomposition of agent will be less.

Normal Combustion Products

Most fires produce copious quantities of heat, smoke, and such toxic combustion products as carbon monoxide, carbon dioxide, and acrolein. In addition, oxygen may be depleted in enclosed spaces to a serious extent. Early detection and rapid extinguishment of a fire will also minimize these hazards.

Approaches to Local Application

Local application systems apply Halon 1301 vapor directly onto the burning surface. NFPA 12A covers local application systems only from an approach standpoint. Since the number of existing systems is still small, the information is basically a guide for manufacturers contemplating introducing these types of systems and for testing laboratories evaluating them.

The intended approach is the extinguishing-quantity-vs.-extinguishing-rate method developed by A. B. Guise [9] and shown in Figure 15.6. The ability of a local application system to extinguish the fire depends on the rate of application of agent to the burning surface. There is a rate, denoted by R_m, at which the fire will never quite be extinguished. As this rate increases, the time required for extinguishment is reduced by an amount greater than the increase in the discharge rate. As a result, the total quantity of agent required for extinguishment (discharge rate times the time to extinguish) is also reduced. At some point, the proportionality reverses. Further increases in discharge rate effect only a small decrease in extinguishing time so that the total quantity required to extinguish begins to increase again. The optimum discharge rate, R_o requires the minimum total agent.

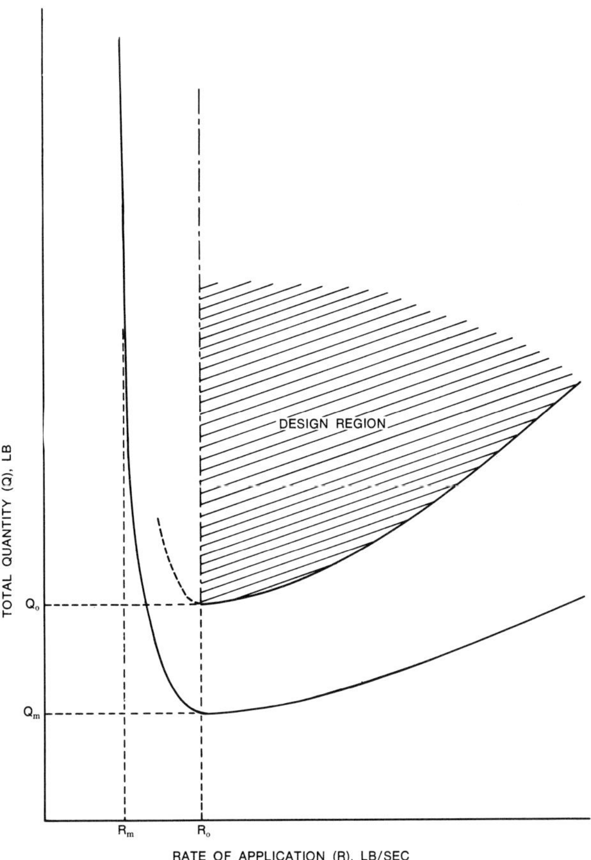

Figure 15.6.

In establishing design parameters for local application nozzles, the discharge rate should not be less than R_o. The total quantity of agent provided should not be less than *twice* the quantity needed to extinguish the fire at any given rate. These two restraints bound the *design region* shaded in Figure 15.6. Each size and type of discharge nozzle could consist of a single curve as shown. An overhead nozzle would produce a family of curves, each representing a specific distance of the nozzle above the flaming surface. Since the experimental program needed to establish these curves is extensive and expensive, there are no local application systems recognized by UL or FM for Halon 1301 service as of this writing. It does seem likely, however, that local application systems will be developed for this agent.

Detector Selection

When a system is designed to be automatic, as most modern systems are, the detectors selected must be considered as carefully as the agent and other system hardware. (For a more complete discussion of detectors, see Chapters 20 and 21.) The following summary is included only for review and applicability to Halon systems.

Detectors should be sensitive enough to fires without being subject to false discharge. (The agent, and hence the cost of recharging the system, are relatively expensive as compared to other gaseous agents.) Also, detectors must be compatible or made to be compatible with the type of power used to operate the container release valve. Heat-actuated detectors (HAD) are pneumatically operated rate-of-temperature-rise devices and are usually compatible only with the same manufacturer's release valve. Simple fixed-temperature thermostats can operate a wide variety of electrical loads but will not differentiate between the high temperatures caused by fire and those from other heat sources. Rate-compensated thermal detectors combine a fixed-temperature thermostat with rate-of-temperature-rise operation to form a reliable, inexpensive and rugged unit.

More sophisticated detectors are available for special hazards. Ionization detectors can detect smoldering-type fires long before substantial quantities of smoke or heat are produced. Optical detectors operate very rapidly to flash fires—both infrared and ultraviolet types are available. Optical smoke detectors are made in both transmission-attenuation and scatter-reflection types. Each type has its own characteristics and should be matched carefully to the hazard.

Improved reliability or performance can often be obtained by using two or more of the same or different types together. A method currently popular is to use an ionization detector for early alarm only and a rate-compensated thermal detector to discharge the system. The sensitive ionization unit allows personnel time to attack the fire with hand equipment. If they are unsuccessful, or if the hazard is unattended, the thermal detector stands ready to discharge the system.

Another current practice is to use cross-zoned ionization detectors. Two of these units must activate alarms simultaneously to discharge the agent. Regardless of the type of detector or detectors used, the alarm signal should be tied into a remote fire control station that is constantly manned. This is good practice for any type of system or agent.

In engineering Halon 1301 systems, special attention must be given to many aspects of the system. This is more important with Halon 1301 systems than with systems using many other types of agents. Manufacturers of fire extinguishing equipment can evaluate individual hazards and recommend the proper system. In addition to their being qualified and familiar with the application of the agent, they will stand behind their work on completion.

References

1. Haessler, W. M., "What You Should Know About Carbon Dioxide Fire Extinguishing Systems," Chapter 14 in this book.

2. Guise, A. B., "A Closer Look At Dry Chemical Fixed-Nozzle Extinguishing Systems," Chapter 16 in this book.

3. Creitz, E. C., "Inhibition Of Diffusion Flames By Methyl Bromide And Trifluoromethyl Bromide Applied To The Fuel And Oxygen Sides Of The Reaction Zone," *J. Res. of Nat'l. Bur. Stds., 65, No. 4.* 1961.

4. NFPA No. 12A-1971, Standard On *Halogenated Fire Extinguishing Agent Systems–Halon 1301,* National Fire Protection Assn., Boston.

5. Miller, M. J., "Extinguishment Of Charcoal, Wood And Paper Fires By Total Flooding With 'Freon' FE 1301/Air And Carbon Dioxide/Air Mixtures," Factory Mutual Research Corp. Serial No. 16234.1, July 28, 1967.

6. R. M. Graziano, Agent, "Hazardous Material Regulations Of The Department Of Transportation, including Specification Of Shipping Containers," *Tariff No. 23,* New York, September, 1969.

7. "Thermodynamic Properties of Du Pont FE 1301 Fire Extinguishant," *Technical Bulletin T-1301.* E. I. du Pont de Nemours & Co. Inc., Freon Products Div., Wilmington.

8. Yao, Cheng and H. F. Smith, "Convective Mass Exchange Between A 'Freon' FE 1301-Air Mixture In An Enclosure And Surrounding Air, Through Openings In Vertical Walls." Factory Mutual Research Corp. Serial No. 16234.1, January, 1968.

9. Guise, A. B., "Extinguishants And Extinguishers," presented to 61st Annual Meeting, National Fire Protection Assn., Los Angeles, May, 1957.

10. Comstock, C. C., F. P. McGrath, S. B. Goldberg, and L. H. Lawson, "An Investigation Of The Toxicity Of Proposed Fire Extinguishing Fluids. Part II—The Approximate Lethal Concentration To Rats By Inhalation Of Vapors For 15 Minutes," *Medical Div. Research Report No. 23.* Chemical Corps, Medical Div. Army Chemical Center, Maryland, October, 1950.

11. Dufour, R.E., "The Life Hazards And Nature Of The Products Formed When Monobromomonochlorodifluoromethane ('Freon' 12B1), Dibromodifluoromethane ('Freon' 12B2), Monobromotrifluoromethane ('Freon' 13B1), Or Dibromotetrafluoroethane ('Freon' 114B2) Extinguishing Agents Are Applied To Fires," *Underwriters' Laboratories Report No. NC-445.* March 10, 1955. Amending letters dated May 1, May 2, and May 31, 1957.

12. Treon, J. F., et al., "The Toxicity Of A Purified Batch Of Monobromotrifluoromethane ('Freon' 13B1), And That Of The Products Of Its Partial, Thermal Decomposition, When Breathed By Experimental Animals." Kettering Laboratory, University of Cincinnati. June 19, 1957.

13. Paulet, G., "Toxicological And Physiopathological Study Of Monobromotrifluoromethane," *Arch. Mal. Prof., 23:341–347, No. 6.* June, 1962.

14. MacFarland, H. N., "Acute Inhalation Exposure—Monkeys, Rabbits, Guinea Pigs And Rats—'Freon' FE 1301—Final Report," Hazleton Laboratories Inc. November 7, 1967.

15. Reinhardt, C. F., "Cardiac Arrhythmias Induced By Epinephrine During Inhalation Of Certain Halogenated Hydrocarbons," *Report No. 14-69.* Haskell Laboratory for Toxicology and Industrial Medicine, E. I. du Pont de Nemours & Co. Inc., January 14, 1969.

16. Van Stee, E. W. and K. C. Back, "Short-Term Inhalation Exposure To Bromotrifluoromethane," *Toxicol. Appl. Pharm., 15:164–174,* July, 1969.

17. Hine, C. H., "Clinical Toxicologic Studies On 'Freon' FE 1301—Report No. 1." Hine Laboratories Inc., January, 1968.

18. Reinhardt, C. F., and G. J. Stopps, "Human Exposures To Bromotrifluoromethane," *Report No. 230–66.* Haskell Laboratory for Toxicology and Industrial Medicine, E. I. du Pont de Nemours & Co. Inc., December 1, 1966.

Dry Chemical Systems

ARTHUR B. GUISE, P.E.

Dry chemical, an important extinguishing agent since 1946, consists of small solid particles suspended in a gas, usually dry nitrogen or carbon dioxide. This ensures flow as well as good streams from nozzles. Before 1950 the information needed to design such two-phase flow systems was not available, but now there are thousands of large and small dry chemcial systems protecting industrial fire hazards as well as restaurant kitchen ranges, hoods and ducts.

Three types of dry chemicals are commonly used in dry chemical systems in the United States and Canada. The major constitutents are sodium bicarbonate, potassium bicarbonate and monoammonium phosphate; the three types are commonly referred to as ordinary, Purple K and multipurpose, respectively. All are effective extinguishing agents for fires in flammable liquids and combustible gases. Only multipurpose dry chemical is an effective agent for fires in ordinary combustibles, such as wood and paper. Being non-conductive, dry chemicals may be used on fires involving live electrical equipment.

Where high effectiveness is desired and powder residues are not objectionable, dry chemical systems offer excellent protection for flammable liquid hazards, especially outdoor hazards where the effectiveness of other agents would be reduced by winds.

Engineered systems are installed on offshore oil drilling platforms, in pump stations, around rolling mills, in oil cellars and paint storerooms. The largest engineered system protects a below-ground-level liquefied natural gas storage tank in New Jersey.

Pre-engineered systems protect smaller hazards: restaurant kitchen range tops, hoods and ductwork; small paint dip tanks, quenching oil tanks and paint

lockers, cotton mill openers, pickers, blenders and associated pneumatic conveyors.

There are places where dry chemical systems should not be used. The powder will not permeate tightly packed combustibles, and so it is not suitable for protecting record vaults, fur storage vaults or similar hazards, which are better protected by permeating gases such as carbon dioxide or Halon 1301.

Dry chemical should not be used where the powder will settle on the contacts of small low-voltage relays or where electronic circuits could be affected.

Large enclosed generators and motors are better protected by flooding with carbon dioxide (see Chapter 14). Large outdoor transformers, switches and circuit breakers are better protected with water spray systems (see Chapter 11) if enough water is available at sufficiently high pressure; otherwise, dry chemical systems give the best protection.

Dry Chemical Characteristics

Dry chemicals contain more than the pulverized chemicals that are their major constituents. Each type contains particulate additives for good flow and is treated with either metal stearates or silicone polymers to repel moisture. They are not toxic under ordinary conditions of use, but a cloud of dry chemical reduces visibility and can cause temporary breathing difficulty; therefore, it is desirable to provide an audible alarm to warn personnel to leave an area before a large system operates.

Being mixtures of solids, dry chemicals produced by different manufacturers vary in particle size and composition and may vary in flow properties and extinguishing effectiveness. Thus equipment is designed to be used with a specific dry chemical. Brand X dry chemical should not be used in Brand Y equipment. To do so might result in a system that would not give the protection for which it was designed.

Multipurpose dry chemical should not be mixed with bicarbonate-base dry chemicals. A chemical reaction takes place that releases carbon dioxide that may hazardously increase the pressure in the container. Such mixtures also produce caking and lumps that upset flow and block nozzles.

Purple K is about twice as effective as ordinary dry chemical [1] but should not be casually substituted in a system designed for the ordinary type. For the same pressure at the nozzles the flow of Purple K may be 10 percent less. The importance of this in a given system is best judged by knowledgeable engineers.

Multipurpose dry chemical is the only extinguishing agent fully effective on fires in ordinary combustibles (Class A), flammable liquids (Class B) and in "live" electrical equipment (Class C). The only limitation on this type for extinguishing Class A fires is that the dry chemical reach all burning surfaces. There is an anomaly: ordinary dry chemical controls fires in cotton fibers, including lint, more rapidly than does multipurpose and, therefore, is used in cotton mills.

Extinguishing Action

In Chapter 14 the extinguishing action of dry chemical is described as primarily the result of removing the hydroxyl (OH^-) radical from the combustion reaction. When ordinary dry chemical was the only type, it was suggested that this action took place by contact of the hydroxyl radical with the cold surface of the particles.[2] A pound of dry chemical has considerable surface area because of the small particle sizes: 1,100 square feet for ordinary, 1,500 square feet for multipurpose and 1,800 square feet for Purple K. Multipurpose is only slightly more effective than ordinary and Purple K is about twice as effective, so it would appear that the extinguishing action is only physical. Tests with sodium and potassium bicarbonates of the same particle size, however, show that potassium bicarbonate is 1.4 times more effective than sodium bicarbonate.[3] The extinguishing action is probably a mixture of chemical and physical, with the former being more important when the rate of application is near the

A 2,000-pound engineered dry-chemical system is for protecting gasoline truck loading racks.

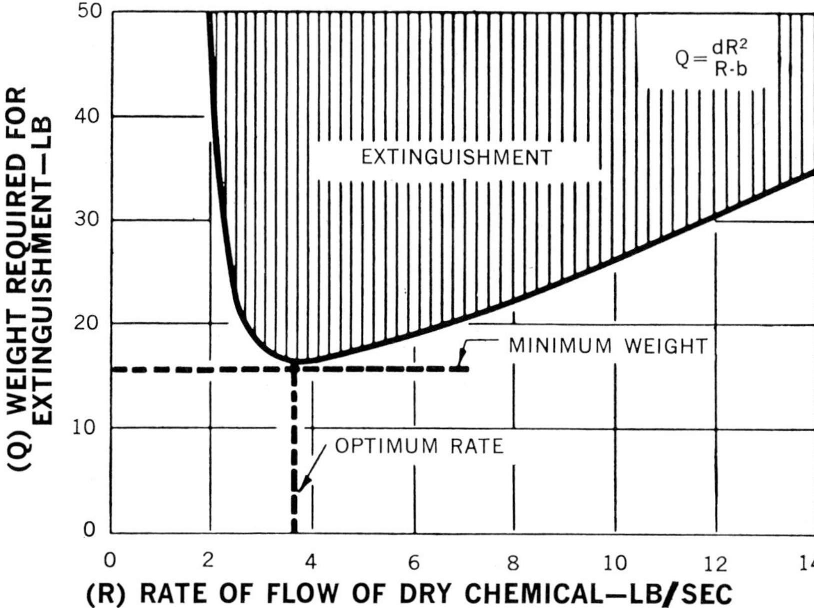

Figure 16.1.

minimum that will extinguish a fire and the latter of greater importance at high rates of application.

The ammonium ion (NH^+_4) has the same electronic configuration as the sodium ion (Na^+). This explains why ammonium phosphate (multipurpose) dry chemical has about the same extinguishing effectiveness as ordinary dry chemical. Multipurpose dry chemical extinguishes glowing embers in wood, etc., because it decomposes when heated to form phosphoric anhydride, which melts to coat the embers.

In Chapter 14 it is shown that carbon dioxide extinguishes by reducing the oxygen concentration. Although dry chemical extinguishes by a different action, the practical methods of use are the same: application of the agents at rates sufficient to build up an extinguishing concentration. Figure 16.1, taken from "A Dry Chemical Extinguishing System," *NFPA Quarterly,* July, 1955, shows that if the application rate is too low there will be no extinguishment.[4] It also shows that there is an optimum rate at which a minimum quantity of agent is required to extinguish. Higher rates extinguish more rapidly but at the cost of more agent. Figure 16.2, taken from "The Chemical Aspects of Fire Extinguishment," *NFPA Quarterly,* April, 1960, shows that this "hook" relationship between rate and quantity is applicable to all agents.[5] The dry chemical used to get the data for Figure 2 was of the ordinary type.

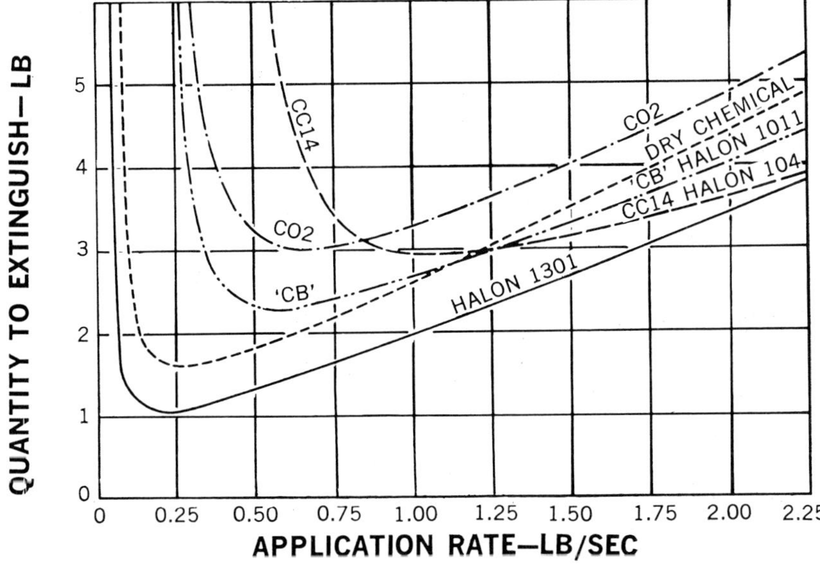

Figure 16.2.

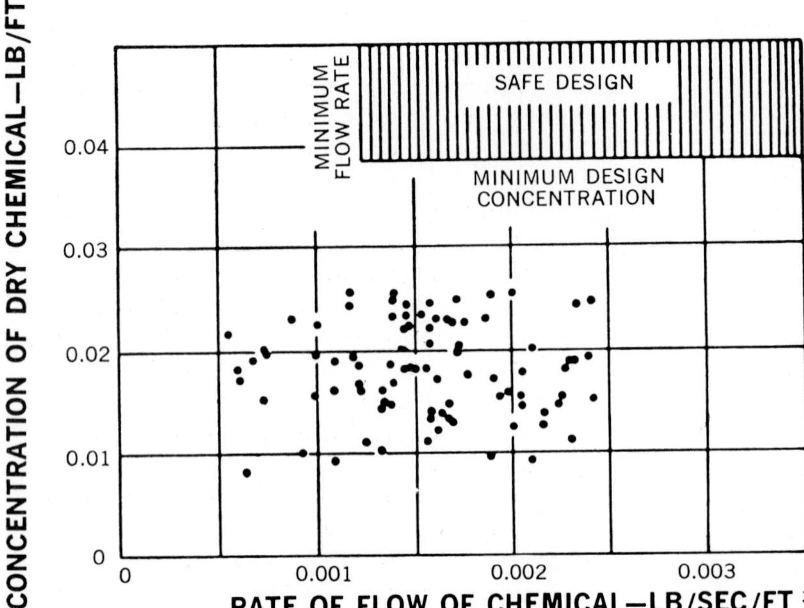

Figure 16.3

For dry chemicals there are no "nose" curves such as for carbon dioxide and Halon 1301. The only published data applicable to total flooding are shown by Figure 16.3, for ordinary dry chemical, and in the UL Fire Protection List. The latter requires a concentration of 0.028 pounds per cubic foot for the extinguishment of Class A, B and C fires by a multipurpose package system.

Although lower concentrations are required for total flooding by dry chemicals, the flooded volume is not inerted for as long a period as with carbon dioxide and Halon 1301. The particles settle out and the concentration is lowered to below an extinguishing level. In practice, when the hazard requires a longer period of inerting, a longer time of flow of dry chemical is used. This counteracts the smaller quantities otherwise needed.

A dry chemical system basically consists of a container for dry chemical, expellant gas for "fluidizing" the powder and expelling it from the container, piping for conveying the dry chemical and nozzles for applying it.

The expellant gas may be stored separately under high pressure in large or small cylinders (the small ones are usually called cartridges), or it may be stored at lower pressures in the same container with the dry chemical. In the latter case, the system is referred to as a "stored-pressure system." Where the gas is stored separately in cylinders, the system is often called a "stored-gas-system."

Systems are further subdivided into "package systems" (also referred to as "pre-engineered systems") and "engineered systems." Package systems have pre-determined flow rates, nozzle pressures and quantities of dry chemical. For these systems, UL or FM prescribe the pipe sizes, maximum and minimum pipe lengths, maximum and minimum number of fittings and number and type of nozzles. The hazards protected by these systems are limited in types and sizes. Engineered systems require individual calculation and design to determine the flow rates, nozzle pressures, quantities of dry chemical, and the number and type of nozzles and their placement in a system.

Package Systems

Most package systems range in container capacity from 2 3/4 pounds to 30 pounds, but a system having a container capacity of 175 pounds has been recently developed.[6] Some are of the stored-pressure type and use nitrogen as the expellant gas; others are of the stored-gas type and use carbon dioxide. The 2 3/4-pound is different because its stored-pressure container is discarded after use and is replaced by a factory-charged container.

Each type has advantages and disadvantages. The stored-pressure type (except for the 2 3/4-pound) requires greater care in charging to make sure the pressurizing gas will not leak away and leave the system inoperative, but it has the advantage that pressure loss can be readily detected with a pressure gauge, which is an integral part of the container. The stored-gas type is more easily charged because permanent gas tightness of the dry chemical container is not so important since the system is under pressure only a few seconds when the system operates. The carbon dioxide cartridge is more easily sealed against gas leakage and is usually factory-charged and tested, but it must be weighed periodically to determine if there is leakage.

The stored-pressure containers can be placed where temperatures are as low as –40° F. without a decrease in extinguishing effectiveness. The cartridges of the stored-gas systems must be at a temperature of 32° F. or higher if the system is to retain effectiveness. Neither type should be located where temperatures are higher than 120° F.

Because the piping and nozzles of a package system are prescribed by UL and FM, no calculations are needed before installation. The limitations in the fire hazards that these systems will protect are determined by fire tests, and the safety factors are undetermined. Consequently, installation must be in accordance with the parameters for piping, nozzles and fire hazards set forth by FM and UL.

An example of the limitations for a specific two nozzle package system:

1. Pipe sizes are 3/4 inch to the reducing tee and 1/2 inch between tee and nozzles.
2. Maximum length of pipe from container to any nozzle is 50 feet.
3. Maximum number of 90-degree elbows between container and any nozzle is six, counting the tee as an elbow.
4. If overhead nozzles are used, no fewer than four elbows must be used (to ensure enough of a pressure drop to reduce the velocity of the streams of dry chemical and thus prevent splashing of the flammable liquid).
5. The lengths of pipe from the reducing tee to the nozzles must be nearly equal, and an equal number of elbows must be in that piping (to get equal flow rates from the nozzles).
6. The fire hazard limitations are: with nozzles at tankside, the maximum area is 40 square feet, length not exceeding 8 feet; with nozzles overhead, the maximum area is 25 feet, length not exceeding 7 feet or width exceeding 3 1/2 feet, and nozzles not nearer than 8 feet or farther than 10 feet above the flammable liquid.

Because the engineering phases of the package system have been completed, an installation can be made without a formal drawing. Manufacturers have prepared installation manuals illustrated with sketches of equipment components and isometric drawings of piping arrangements. Such manuals form a part of the kit sold by the manufacturer, are given a part number, and are examined by UL.

With the aid of the installation manual, a competent pipe fitter or mechanic can correctly install a package system. However, the directions for piping are sometimes not followed. Because he does not know—and sometimes doubts even if he does know—why dry chemical is piped in a certain way, the installer will sometimes pipe the dry chemical as if it were air or water. This can result in an ineffective system.

Though package systems, in single units, are limited to fire hazards of specific maximum size, they can be used in multiples of up to five systems, simultaneously operated, to protect larger hazards. Each individual system still has the original limitations imposed by UL and FM. Two methods are used for obtaining simultaneous operation: mechanical, which requires that the containers be adjacent and grouped in line, and pneumatic, which allows the containers to be separated but all must be within 100 feet of the actuating source, a carbon dioxide cartridge.

Though the package systems were developed as protection for small industrial hazards, they are commonly used in restaurant kitchens to extinguish grease fires on range tops and in the hood and duct ventilating systems. At first, such protection was supplied by simple adaptations of the industrial systems, but now the kitchen systems are highly specialized.

It was stated above that the safety factors of package systems are undetermined. That does not mean that there are no safety factors; it means they have not been publicized. The safety factors, if any, are known by the manufacturers, UL and FM. If all actual fire hazards duplicated the "hazards" used by UL and FM in setting the limitations of a system, the fire protection engineer would have no problem. As it is, however, he is in the position of a mechanical engineer trying to design a structure or pressure vessel without knowing the allowable stress of the metal he proposes to use.

The safety factors of engineered systems have been published. Using that information and the known UL limitations on a five-unit multiple package system, the safety factors for that one particular system can be estimated as shown in Table 16.1. From the table, it can be estimated that this particular package system

Table 16.1. Pre-Engineered and Engineered Systems Comparison

	Flammable Liquid Area sq ft	Room Flooding Volume cu ft
5 30-lb, pre-engineered, as listed	200	5,000
150 lb, engineered, as listed with safety factor	172	3,900
150 lb, engineered, no safety factor	215	7,500

has a small safety factor for tanks of flammable liquids and has a good safety factor for total flooding.

Correct installation of package systems becomes imperative if the fire hazard approaches the upper allowable limits of size. One way to make sure that there is a safety factor is to use a multiple-unit system with one more unit than if the limitations set by UL and FM were to be applied.

Engineered System Design

Engineered systems have dry chemical container capacities ranging from 150 pounds to 2,000 pounds. Like package systems, these can be arranged for simultaneous operation of multiple units for the protection of very large fire hazards. The expellant gas is nitrogen stored in DOT cylinders from which it is supplied to the dry chemical container after being reduced in pressure to 220 psi by means of pressure regulators. The pressure in the cylinders can be checked by pressure gauges to ensure that there is gas to effectively operate the system.

It was pointed out earlier that to extinguish a given fire, an agent must be applied more rapidly than the minimum rate determined by tests and that the quantity of agent must exceed a minimum also determined by tests. For engineered systems using ordinary dry chemical, those minimums were determined by UL and FM after more than 1,600 fire tests with gasoline. In designing a system, a safety factor is applied by using an application rate of not less than twice the failure rate and by using at least twice the minimum quantity of dry chemical. The reason for the safety factor is that seldom are industrial hazards as simple as test "hazards."

The first step in designing an engineered system is to select the application rate and the quantity of dry chemical. The next step is to select the nozzles and decide where and how they should be placed. The third step is to lay out the piping, and the final step is to calculate pipe sizes so that pressure drop is not excessive. The relationship between flow rate and container outlet pressure, pressure drop in pipe and fittings, and nozzle pressure have been determined experimentally and reduced to empirical equations.

The advantage of an engineered system is that there is greater flexibility in design. There are no limitations on pipe size or length except those imposed by pressure drop. There are no limitations on the number of nozzles. The only limitation is that dry chemical must be applied on a fire in an effective manner—with a safety factor. The design engineer must be knowledgeable in the flow characteristics of dry chemical and familiar with the practical aspects of industrial hazards. The installation

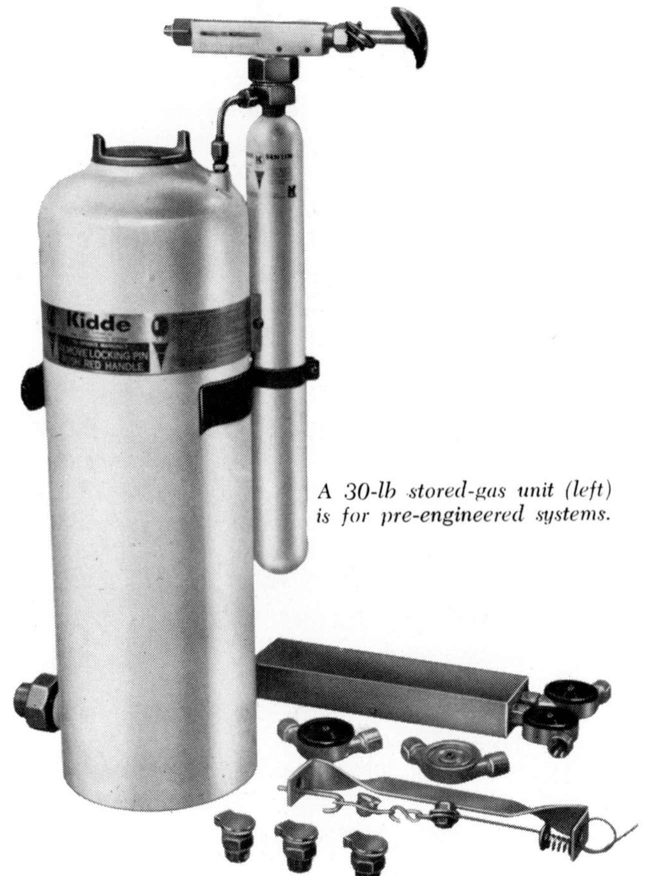

A 30-lb stored-gas unit (left) is for pre-engineered systems.

(Above) Arranged for multiple operation, 30-lb pressure units are for pre-engineered systems.

*The 150-lb dry-chemical unit is
for engineered systems (below).*

should be supervised by an engineer or competent technician and finally inspected by an engineer or technician experienced with dry chemical systems.

The flow of dry chemical through pipe and fittings is two-phase: a suspension of solid particles in gas. Information applicable to pneumatic conveying of solids such as grain or powdered coal is not applicable to dry chemical systems because the particle sizes of dry chemicals are much smaller, the solids-to-gas ratio much greater, and the mass flow rates are greater. The solids-to-gas ratio can vary with equipment design, and particle sizes vary according to the manufacturer, even for the same type of dry chemical. Experimental information on flow characteristics is therefore related to a dry chemical and to the equipment from which it is discharged. Applying the same data, without tests, to other equipment and different dry chemical could result in the design of a system which would be ineffective.

Pressure drop in pipe can be expressed as $\Delta P = kR^n/d^5$ where k and n are constants that will vary with dry chemicals and equipment design. R is the mass flow rate, and d is the inside diameter of pipe. The expression is not exact but gives practically usable results.

In attempting to use large pipe sizes to reduce pressure drop, care must be taken to avoid surging. In long, straight runs of pipe, particles settle to the bottom of the pipe if the velocity is too low to cause turbulent flow. As the solids settle, the stream becomes "gassy" and ineffective. Then as the pipe becomes constricted by the settled solids, the velocity increases so that the gas stream picks up the settled solids and the combined stream has higher than normal solids. The velocity then decreases and the process repeats itself.

The cure for surging is to use flow rates that introduce enough turbulence to prevent settling. For one system and ordinary dry chemical the critical minimum flow rates range from 1 pound per second for 1/2-inch pipe to 45 pounds per second for 4-inch pipe. Elbows and tees, but not couplings, cause enough turbulence in a line to prevent surging.

Dry chemical streams have other peculiarities. For the same mass flow rate, the pressure drop in straight pipe is only about two-thirds that of water. However, in elbows, bends and tees off the run the pressure drop with dry chemical is much greater than with water. Further, with water the equivalent pipe length increases as the pipe size decreases; with dry chemical, the equivalent pipe length decreases as pipe size increases. There is no advantage in using large radius bends instead of elbows.

As dry chemical streams are redirected by elbows, bends or tees, the solids are thrown to the outside of the bend and tend to remain separated from the gas until turbulence causes remixing. Because of this, streams are always split by entering a tee from off the run so that the subdivided streams retain the original solids-to-gas ratio. Subdividing by taking a subsidiary stream off the run of a tee results in a high solids-to-gas ratio in the stream on the run and a "gassy," ineffective stream off the run.

Piping is further complicated if an elbow or bend is in the piping to the tee and is in the same plane as the tee. The solids are thrown to the outside curve of the bend and, if the solids and gas are not remixed before entering the tee, one stream leaving the tee will have a high solids-to-gas ratio and the other will be gassy.

There are two methods used to remix before the dry chemical enters the tee. One is to use a pipe length of at least twenty pipe diameters between the bend and the tee, and the other is to use a special fitting having an inside diameter smaller than the pipe diameter, with a good entrance and exit coefficient, between the bend and the tee (see Figure 16.4).

The foregoing implies that a high solids-to-gas ratio is always effective. This is not true because a higher than normal solids-to-gas ratio will not only reduce stream range but the particles also may not be dispersed into an effective cloud of dry chemical. The system has been designed to expel the dry chemi-

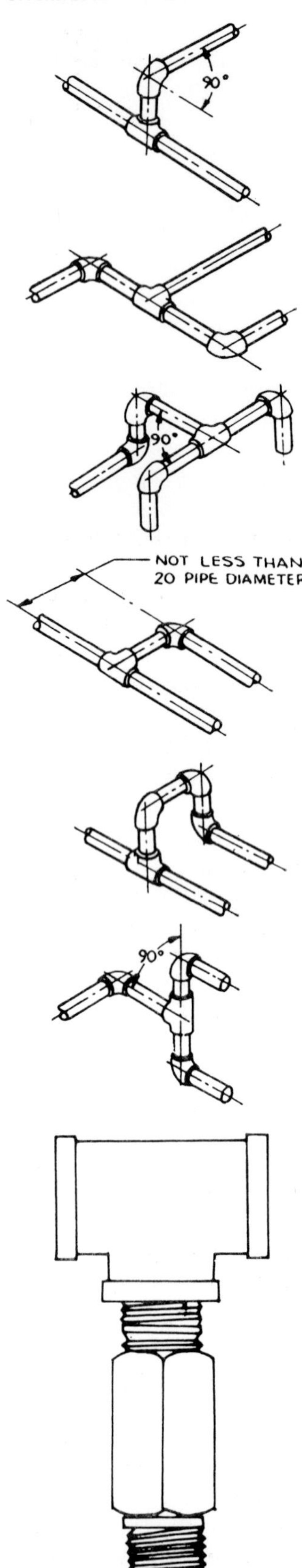

Figure 16.4.

cal in a specific solids-to-gas ratio, and piping must follow good practice for dry chemical systems or the streams from the nozzles will vary in effectiveness.

Further information on dry chemical systems can be obtained from the *Standard for Dry Chemical Extinguishing Systems,* NFPA No. 17, which may be obtained from the National Fire Protection Assn., 470 Atlantic Ave., Boston, Mass. 02110, for 75 cents. Information can be obtained from manufacturers listed in the Fire Protection Equipment List of the Underwriters Laboratories, Inc. or the Approved Equipment list of the Factory Mutual Engineering Co.

References

1. Guise, A. B., "Potassium Bicarbonate-base Dry Chemical," *NFPA Qtly,* July, 1962, pp. 2–9.

2. Guise, A. B., "Dry Chemical Extinguishment Development," *Proc. Symp. on Fire Extinguishment Res. and Eng'g,* U.S. Naval Civil Eng'g Res. and Evaluation Lab., Port Hueneme, Cal., November 1954, pp. 363–370.

3. Vanpee, M. and others, "Inhibition of Afterburning by Metal Compounds," *Progress in Astronautics and Aeronautics,* Vol. 15, Academic Press, 1964, pp. 419–448.

4. Guise, A. B. and Lindlof, J. A., "A Dry Chemical Extinguishing System," *NFPA Qtly,* July, 1955, pp. 3–11.

5. Guise, A. B., "The Chemical Aspects of Fire Extinguishment," *NFPA Qtly,* April, 1960, pp. 330–336.

6. Cholin, R., "The Development of the Safety First Automatic System Model 250," *Fire Technology,* September, 1969, pp. 14–19.

7. Standard for Dry Chemical Extinguishing Systems, NFPA No. 17, National Fire Protection Assn., Boston, Mass.

Fire Extinguishers

MARSHALL E. PETERSEN, P.E.

A great deal of attention has been given to the first five minutes of fire. In its early stages, almost any fire can be extinguished by means of the proper type of portable* extinguisher. But will this extinguisher be the proper type, in place and ready for use? Primary responsibility for the selection and placement of extinguishers frequently lies with the mechanical engineer and the owner of a property.

There are various types, styles, and sizes of portable extinguishers. Each fills a particular need in fire protection. Their necessity and importance is confirmed by the fact that fire insurance credits are given for conformance to recognized standards, regardless of what other fire equipment (automatic sprinklers, hose lines, etc.) might exist in an occupancy.

The National Fire Protection Assn.'s Standard No. 10, *Portable Fire Extinguishers,* classifies extinguishers according to the kinds of fires they are used on. Since this standard has been adopted by the Occupational Safety and Health Administration, it is advisable that a copy be obtained as a guide for selecting and installing fire extinguishers. (It is available from NFPA, 470 Atlantic Ave., Boston 02110.)

Class A extinguishers are suitable for use on fires in ordinary combustibles, such as wood, paper, rubber and many plastics. They are used where the quenching, cooling ability of water or of solutions containing large percentages of water are most effective. Extinguishers rated for Class A hazards are water and loaded-stream. Multipurpose dry-chemical types are also suitable because of their special ability to retard com-

bustion. (Figure 17.1 shows the various types of extinguishers available.)

Class B extinguishers are used on fires in petroleum products, flammable gases and other flammable materials, such as paints, thinners and solvents. They are effective where an oxygen-exclusion or flame-interruption effect is essential. Extinguishers rated for Class B hazards are carbon-dioxide, ordinary dry-chemical, multipurpose dry-chemical, and to a limited extent, loaded-stream.

Class C extinguishers are for fires in energized electrical equipment and wiring, where the dielectric conductivity of the extinguishing agent is of importance. Water or water-solution extinguishers, of course, cannot be used on electrical fires because of the shock potential.

Class D extinguishers are used on fires in combustible metals, such as magnesium, titanium, zirconium, sodium and potassium. Class D extinguishers are designated "dry powder" extinguishers and contain such specially treated agents as sodium chloride or graphite.

Quality in a fire extinguisher is assured when it bears the label of a recognized testing laboratory, such as Underwriters Laboratories, Inc., Underwriters Laboratories of Canada or Factory Mutual Laboratories.

UL standards are based on the NFPA classifications and principles of fire extinguishment. The UL standards also consider fire extinguishment potentials determined by physical testing. These standards verify that different types of fires can be best extinguished by different methods.

In addition to determining the degree of suitability for a particular class of fire, UL also conducts tests to determine the strength of parts and assemblies, corrosion-resistance of the materials used and the operating characteristics under normal and abnormal conditions

---

*"Portable" is applied to manual equipment used on small fires or used in the time between discovery of fire and the functioning of automatic equipment or arrival of professional fire fighters.

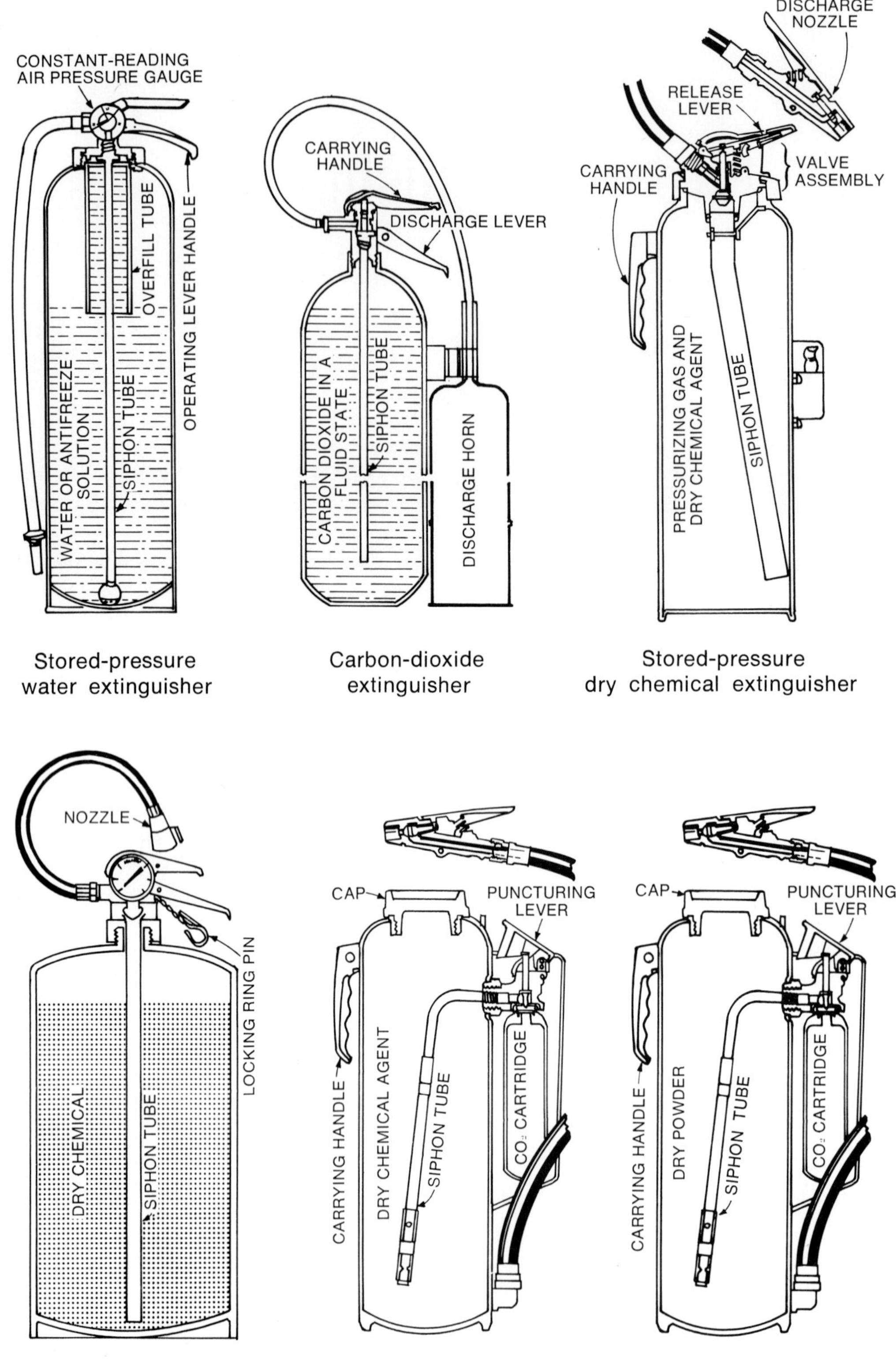

Stored-pressure
water extinguisher

Carbon-dioxide
extinguisher

Stored-pressure
dry chemical extinguisher

Stored-pressure
dry chemical extinguisher

Cartridge-operated
dry chemical extinguisher

Dry powder
extinguisher

(see How UL Tests Extinguishers at the end of this chapter).

What UL Numerals Mean

A UL label not only guarantees workmanship but also gives the relative fire extinguishing capabilities of the extinguisher and tells what type of fire it will effectively extinguish. The standard fire extinguisher classification consists of a numeral and a letter. It appears on the label affixed to all fire extinguishers labeled by Underwriters Laboratories, Inc. and Underwriters Laboratories of Canada.

The UL numerals and letters denote the following:

Class A extinguishers. The numeral indicates the approximate, relative fire extinguishing potential of various sizes of Class A fire extinguishers. For example, a 4-A extinguisher can be expected to extinguish approximately twice as much fire as a 2-A extinguisher.

Class B extinguishers. The numeral indicates the approximate, relative fire extinguishing potential of various sizes of Class B fire extinguishers. In addition, the numeral is an approximate indication of the square foot area of deep-layer flammable liquid fire that an inexperienced operator can extinguish—e.g., a 10-B unit can be expected to extinguish ten square feet of deep-layer flammable liquid fire.

Class C extinguishers. No numeral is used because Class C fires are essentially either Class A or B fires that involve energized electrical wiring and equipment. The size of the Class C extinguishers specified should be commensurate with the size and extent of the associated class A, B or D fire hazard, considering that a potential fire must be covered or blanketed by the Class C extinguishing agent for effective extinguishment.

In addition to the letter C, other letters (A and/or B) and numerals refer to the classes and size of fires on which use of the particular extinguisher is approved for most effective extinguishment.

Class D extinguishers. No numeral is used for this class. The relative effectiveness of these extinguishers for use on *specific* combustible metal fires is detailed on the extinguisher nameplate.

Identifying, Locating Extinguishers

Extinguishers and extinguisher locations should be marked to indicate the suitability of the extinguisher for a particular class of fire. The recommendations for marking extinguishers are contained in NFPA 10 (see Figure 17.2).

Markings should be applied by delcalcomanias, painting or similar methods having at least equivalent legibility and durability. Marking applied directly on the extinguisher should be located on the front of the shell above or below the extinguisher nameplate. It should be legible from a distance of at least three feet.

Markings applied to wall panels, etc., in the vicinity of extinguishers should be of a size and form easily read from a distance of twenty-five feet.

Extinguishers must be located close to likely hazards—but not so close that they would be damaged or isolated from the fire. If possible, they should be located along normal paths of egress from the building. Where highly combustible material is stored in small rooms or enclosed spaces, extinguishers should be located outside the door—never inside the room, where they might become inaccessible.

Extinguishers must not be blocked or hidden by stock, finished material or machines. They should be located or hung where they will not be damaged by trucks, cranes and harmful operations, or corroded by chemical processes. They must be placed where they will not obstruct aisles or injure passersby.

Extinguisher locations should be conspicuous. For example, if an extinguisher hangs on a large column or post, a distinguishing red band can be painted around the post. Also, it is advisable to post large signs directing attention to the extinguishers. Extinguishers should be kept clean and should not be painted in any way that would camouflage them or obscure labels and markings. Extinguishers having a gross weight of forty pounds or less should be installed so that the top of the unit is not more than five feet above the floor. Hand-portable extinguishers that weigh more than forty pounds should be installed so the top of the unit is not more than three-and-one-half feet above the floor. If women may use them, consideration should be given to mounting the extinguishers at a lower height. Extinguishers should not, however, be mounted closer than six inches from the floor.

Improper mounting of fire extinguishers is commonly cited as an OSHA violation, since it is easily detected during compliance inspections.

Number and Type

Factors determining the number of extinguishers needed to protect a property include the area and arrangement of the building or occupancy, the severity of the hazard, the anticipated classes of fires, and the distances to be traveled to reach extinguishers.

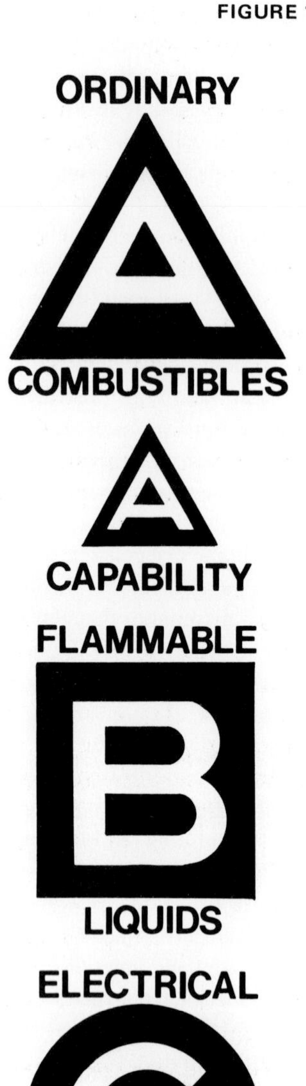

Labels extinguishers containing water or water-based agents, or multipurpose chemical agents (water pump tanks, pressurized water, soda acid, foam, and monoammonium phosphate or diammonium phosphate base agents). Color: green.

Use on fires in wood, paper, textiles, trash.

Labels extinguishers containing: multipurpose dry chemical (A:B:C)—but not rated for Class A fires. These are small extinguishers charged with A:B:C dry chemical but have insufficient effectiveness to earn the minimum 1A rating, even though they have some value in extinguishing small Class A fires. (They are rated, however, for Class B and C fires.) Color: green.

Labels extinguishers containing smothering or flame-interrupting chemicals (CO_2, sodium and potassium bicarbonate base dry chemicals, multipurpose chemicals, bromotrifluoromethane, carbon tetrachloride, chlorobromomethane and foam). Color: red.

Use on fires in flammable liquids such as gasoline, paint, solvents, oil, tar.

Labels extinguishers containing nonconducting extinguishants (CO_2, sodium and potassium bicarbonate base dry chemicals, multipurpose chemicals, bromotrifluoromethane, carbon tetrachloride and chlorobromethane). Color: blue.

Use on fires or near electrical equipment.

For labeling containers of special dry powder extinguishing agents capable of smothering fires in combustible metals (graphite base agents, graphite-phosphorus compounds and sodium chloride base powders). Color: yellow.

Use on fires in combustible metals such as sodium, titanium, uranium, zirconium, lithium, magnesium, sodium-potassium alloys.

NFPA 10 grades rooms or areas generally as light hazard, ordinary hazard or extra hazard.

A light-hazard location is one where the amount of combustibles or flammable liquids is such that small fires may be expected. Such locations usually include offices, school rooms, churches, assembly halls, telephone exchanges, etc.

An ordinary-hazard location is where the amount of combustibles or flammable liquids present is such that moderate size fires may be expected. These locations usually include mercantile storage and display areas, auto showrooms, parking garages, light manufacturing operations, warehouses not classified as extra hazard, and school-shop areas.

In extra-hazard locations the amount of combustibles or flammable liquids present is such that fires of severe magnitude may be expected. These include woodworking areas, auto repair shops, aircraft servicing operations, warehouses with high-piled combustibles, and such processes as flammable liquid handling, painting and dipping.

The relative fire hazard of each floor's or area's occupancy, the nature of any anticipated fires and protection for special hazards determines the minimum number and type of portable extinguishers specified. Extinguishers suitable for Class A hazards are specified according to the classification of occupancy: light, ordinary or extra hazard (summarized in Table 17.1).

Extinguisher requirements for Class B protection (special-hazard areas, such as laboratories, restaurant kitchens and paint spray booths) are in addition to the requirements of extinguishers for Class A protection, except where the total area under consideration presents only Class B hazards. The requirements for fire extinguisher size and placement for Class B fires are shown in Table 17.2.

Extinguishers with Class C ratings are required where energized electrical equipment that would require a nonconducting extinguishing media may be encountered. This will include a fire either directly involving or adjacent to electrical equipment. Since the fire itself is a Class A or Class B hazard, the extinguishers are sized and located on the base of the anticipated Class A or B hazard. In all cases, scattered or widely separated hazards should be individually protected if the specific travel distances are exceeded.

Extinguishers used for Class D protection and other special fires are installed according to the size and type of special hazard. The type and quantity of combustible material and its physical form are determining factors that must be considered when selecting the proper type of agent and the method of application.

Tips on Selection

Choosing the right extinguisher is extremely important. Too frequently cost is given more consideration than is adequate protection. Although suitable types and sizes are stipulated by NFPA 10, there is a certain flexibility that can be used in extinguisher selection. For example, if a certain condition calls for protection by a 10-B unit, either a five-pound ordinary dry-chemical (sodium bicarbonate), a fifteen-pound carbon-dioxide, a two-and-one-half-pound potassium-bicarbonate dry-chemical or a four-and-one-half-pound multipurpose dry-chemical extinguisher could be used.

The relative advantages and disadvantages of each extinguisher should be considered with respect to all conditions where used. It is also wise to investigate the merits of particular extinguishers available on the market. There are ordinary dry-chemical extinguishers

Table 17.1. Where to Specify Class A Extinguishers

Basic minimum extinguisher rating for area specified	Maximum travel distances to extinguishers	Areas to be protected per extinguisher		
		Light hazard occupancy	Ordinary hazard occupancy	Extra hazard occupancy
1A	75 ft	3,000 sq ft	Not permitted except as specified in Paragraph 4120	Not permitted except as specified in Paragraph 4120
2A	75 ft	6,000 sq ft	3,000 sq ft	Not permitted except as specified in Paragraph 4120
3A	75 ft	9,000 sq ft	4,500 sq ft	3,000 sq ft
4A	75 ft	11,250 sq ft*	6,000 sq ft	4,000 sq ft
6A	75 ft	11,250 sq ft*	9,000 sq ft	6,000 sq ft
10A	75 ft	11,250 sq ft*	11,250 sq ft*	9,000 sq ft
20A	75 ft	11,250 sq ft*	11,250 sq ft*	11,250 sq ft*
40A	75 ft	11,250 sq ft*	11,250 sq ft*	11,250 sq ft*

*11,250 sq ft is considered a practical limit.

| | For extinguishers labeled prior to June 1, 1969 | |
Type of hazard	Basic minimum extinguisher rating	Maximum travel distance to extinguishers
Light	4B	50 ft
Ordinary	8B	50 ft
Extra	12B	50 ft

| | For extinguishers labeled after June 1, 1969 | |
Type of hazard	Basic minimum extinguisher rating	Maximum travel distance to extinguishers
Light	5B	30 ft
	10B	50 ft
Ordinary	10B	30 ft
	20B	50 ft
Extra	20B	30 ft
	40B	50 ft

(sodium bicarbonate) of the ten-pound size that have a 10-B:C or a 20-B:C or a 30-B:C rating.

Operating characteristics that make one type of portable extinguisher suitable for certain fire hazards may make the same type dangerous for others. Additional factors include design and operating features, ease of maintenance and availability of repair service. Safety and health hazards involved in extinguisher maintenance and use must also be considered.

In general, the types of extinguishers available are:

1. Class A: stored-pressure water, loaded-stream and multipurpose dry-chemical.
2. Class A and B: loaded-stream.
3. Class B and C: carbon-dioxide and ordinary dry-chemical.
4. Class A, B and C: multipurpose dry-chemical.

Class A Extinguishers

The most common and suitable extinguisher for Class A protection is the stored-pressure water extinguisher. It is available in a two-and-one-half-gallon size and is rated by UL at 2-A. It can be installed in all light- or ordinary-hazard occupancies but not in extra-hazard occupancies.

Extinguishers of this type must be located throughout an occupancy so that no one will be more than seventy-five feet from the nearest unit. They should also be located, as much as possible, near exits and other normal paths of travel. Placement of the extinguisher should be on a hanger or in a cabinet so that the top of the extinguisher is no more than three-and-one-half feet above the floor.

Class A extinguishers are intended for installation in areas where the ambient temperature is between 40° F. and 120° F. Where temperatures below 40° F. are anticipated, an antifreeze charge available from the extinguisher manufacturer should be used.

Class A:B Extinguishers

The only extinguisher available in this category is loaded-stream. It is identical to the stored-pressure water type except that in place of water the extinguisher contains two-and-one-half gallons of loaded-stream extinguishing agent. This liquid is an alkali-metal-salt solution that will not freeze at temperatures of –40° F. Depending on the manufacturer, these extinguishers are available with ratings of 2-A:1-B (for light- or ordinary-hazard occupancies), 3-A:1-B (for all classes of occupancy). Maximum travel distance to the nearest unit is seventy-five feet.

Class B:C Extinguishers

Two extinguishers are available for this category of fire—carbon dioxide and several types of ordinary dry-chemical. Carbon-dioxide extinguishers are available in hand-portable models ranging from two-and-one-half to twenty pounds with corresponding UL ratings of 1- to 10-B:C. There are wheeled extinguishers in capacities from fifty to one hundred pounds with ratings from 10- to 40-B:C. When carbon dioxide is discharged onto a fire its gaseous state leaves no residue. Therefore, this type of extinguisher is generally used to protect electronic equipment, laboratories, computer equipment, kitchens and other hazards or areas where a "clean" extinguishing agent is preferred.

The main disadvantage of the carbon-dioxide extinguisher is that it cannot be used efficiently outdoors or where there are strong air currents. This is so because of its short discharge range (three to eight feet) and the gaseous nature of the agent. Since carbon dioxide will not support life, it is also important not to install large capacity units in unventilated spaces, such as small rooms or confined areas.

Ordinary dry-chemical extinguishers come in several different models (cartridge-operated or stored-pressure) and with different types of extinguishing agent (sodium-bicarbonate base, potassium-bicarbonate base or potassium-chloride base). Hand-portable models range in capacity from one to thirty pounds with corresponding UL rates of 1-B:C to 80- B:C. Wheeled extinguishers have capacities ranging from fifty to 350 pounds and ratings from 40-B:C to 320-B:C.

The decision to specify a cartridge-operated model instead of a stored-pressure model for a particular installation generally rests on the convenience of self-recharging and maintenance. Both models do an equivalent job of fire extinguishment, and since few owners do their own recharging, either is satisfactory.

The type of agent, however, makes a difference. The most common and universal extinguisher uses an agent with a sodium-bicarbonate base. If a more powerful agent or a smaller capacity extinguisher for a given rating is desired, an agent with a potassium-bicarbonate base (Purple K) can be used. It costs about twice as much per pound and will extinguish about twice as much fire on a pound-per-pound basis. Some manufacturers also have a potassium-chloride-base agent that is slightly less expensive but has about the same fire-killing power as the potassium-bicarbonate-base agent. The potassium-chloride-base agent, however, is acidic and causes corrosion when discharged on a fire that involves metallic materials. The bicarbonate-base agents will not cause corrosion. Ordinary dry-chemical agents leave a white or purple powdery residue that can be cleaned up by sweeping or with a vacuum.

Class A:B:C Extinguishers

This type of extinguisher is referred to as "multi-purpose" dry-chemical. It comes in both cartridge-operated and stored-pressure models. Hand extinguishers range in capacity from four to thirty pounds and have corresponding UL ratings from 1- to 10-A and 10- to 60-B:C. Wheeled units have capacities of fifty to 300 pounds and ratings of 20- to 40-A and 60- to 240-B:C. Certain smaller capacity models charged with multi-purpose dry-chemical are rated on Class B and C fires only when the amount of agent is insufficient to earn the minimum 1-A rating. In such cases, the Class A symbol is printed on the nameplate along with the word "capability."

The multipurpose agent consists primarily of ammonium phosphate, which is acidic and will corrode metals. When discharged into a fire the agent becomes soft and sticky and will easily adhere to hot surfaces. After the burning materials cool, the agent will harden. Generally, cleaning is more difficult than with ordinary dry-chemical agents. (This type of extinguisher is not usually recommended for protection of electrical equipment and machinery.)

Class D Extinguishers

Though these extinguishers are referred to as "dry powder," they should not be confused with the term "dry chemical" used for Class B and C fires. Class D extinguishers come in thirty-pound hand-portable models and 350-pound wheeled models. They are available with several different agents—Met-L-X (sodium chloride and tricalcium phosphate), Lith-X (granular graphite) and Met-L-Kyl (bicarbonate base with activated absorbent). It is important to note that if the combus-

tible metal on fire and the extinguishing agent to be used are not compatible, a violent reaction may result. When specifying fire extinguishers for Class D protection (combustible metal fires), the mechanical engineer should consult a fire protection engineer, chemical engineer or a manufacturer of Class D extinguishers.

Locating B:C and A:B:C Models

If a hand-portable Class B:C and A:B:C extinguisher weighs less than forty pounds, the top of the extinguisher should be no more than five feet above the floor. If it weighs more, the top should be no more than three-and-one-half feet from the floor.

If extinguishers in this category are to be in areas where the temperature may go below –40° F. or above 120° F., the extinguisher manufacturer should be consulted.

The type of hazard—light, ordinary, or extra—determines the *minimum* UL rating and the *maximum* travel distance to the extinguisher for Class B protection.

For flammable liquid hazards of appreciable depth (greater than one-quarter inch), as in dip or quench tanks, Class B fire extinguishers should be specified on the basis of one numeral unit of Class B extinguishing potential per square foot of flammable liquid surface of the largest tank hazard within the area.

When protection is sought for flammable liquid in appreciable depth, and when the liquid surface area is in excess of twenty square feet, the protection requirements must be based on an evaluation of the extent of the hazard. Consideration should be given to specifying fixed protection systems and wheeled extinguishers. Portable-extinguisher protection for such hazards should be generally restricted to plants having trained fire brigades.

Class A Number, Placement

The number of Class A extinguishers needed for a given area can be estimated by using Table 17.1. For example, if two-and-one-half-gallon stored-pressure water extinguishers (which UL rates 2-A) are to be specified for an ordinary-hazard occupancy, the table indicates each extinguisher will protect an area of 3,000 square feet. Therefore, if the floor area is 150 feet × 250 feet (37,500 square feet), dividing by 3,000 will give 12 1/2, or 13 units (all fractions should be counted as an additional unit). For a light-hazard occupancy, it would be 37,500 divided by 6,000, or seven units.

The maximum allowable travel distance to Class A

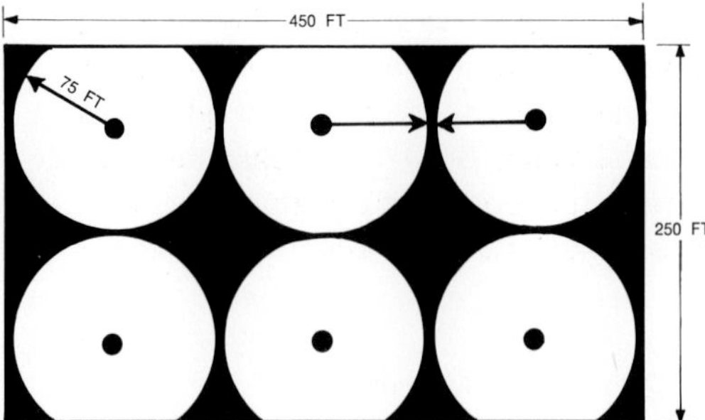

Wrong: inadequate number of Class A extinguishers.

Figure 17.3.

extinguishers is seventy-five feet. Arranging extinguishers at intervals of 150 feet as shown in Figure 17.3 will not conform to this requirement because persons in the dark areas would need to travel more than 75 feet to reach the nearest unit. A better arrangement is shown in Figure 17.4.

Considerable freedom of placement is allowed if the arrangement has no void spaces (areas more than seventy-five feet away from the nearest unit). Taking the previous example of 37,500 square feet that required thirteen units, we can place units as shown in Figure 17.5 or in any other pattern that will give complete coverage. In actual installations travel distances around partitions, machinery, walls, stock and other obstructions must be taken into account when measuring the seventy-five feet.

Class B: How Many and Where?

Class B extinguishers can be installed in two ways: as spot-hazard protection (dip tank, spray booth, etc.) or as protection for an area totally comprised of Class B hazards.

The size and travel distance for Class B extinguishers are given in Table 17.2. The degree of hazard gives

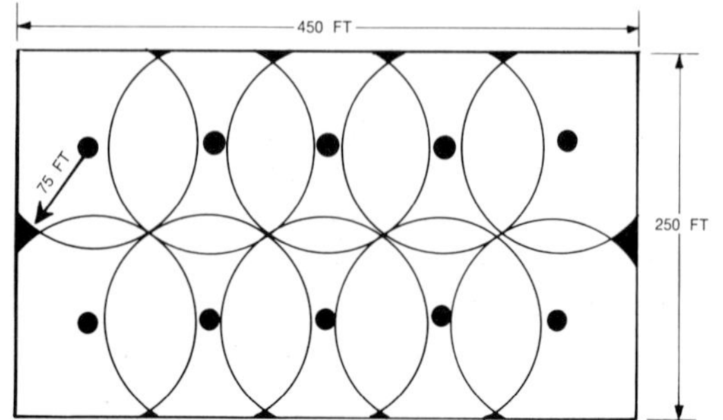

Right: proper number of Class A extinguishers; none is more than 75 ft from any point. (Dark areas are beyond the 75-ft travel distance; however, it has been common practice to ignore such small areas, especially when they are at exterior walls.)

Figure 17.4.

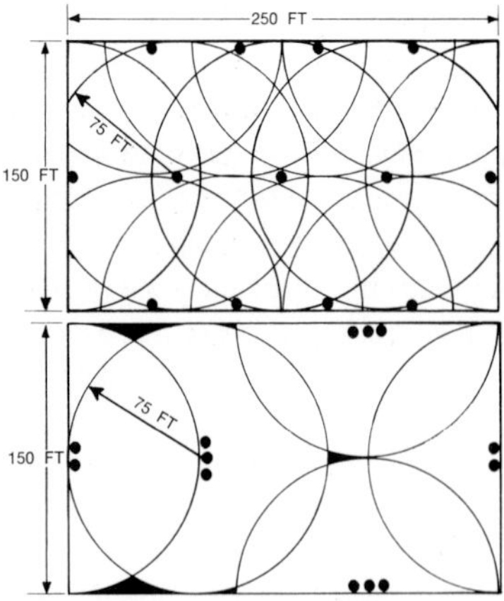

This size area requires 13 Class A extinguishers. Shown are two of the many ways the extinguishers can be placed to provide coverage conforming to the 75-ft travel distance requirement. The second example illustrates group placement. Although in the diagram, the grouping has resulted in some areas beyond the 75-ft travel distance, these areas can be easily eliminated by spacing the individual extinguishers farther apart.

Figure 17.5.

guidance on the size of unit to select and its equivalent travel distance.

For spot-hazard protection, the travel distance figure in Table 17.2 is not the required distance between hazard and extinguisher. The general practice is to locate units for spot-hazard protection closer to the hazard—but not so close that they will be inaccessible if there is a fire.

When two or more hazards are to be protected, a single extinguisher may be used if the hazards are not more than one hundred feet apart and if it is unlikely that a fire in one would easily spread to the other. In selecting an extinguisher to protect several spot hazards, the size should be based on that required by the largest hazard.

When an occupancy or area is entirely a Class B hazard, extinguishers should be specified according to the degree of hazard (light, ordinary or extra) and placed using the appropriate travel distance figures. For example, a paint-storage room judged to be an ordinary hazard could be protected by 10-B extinguishers with a maximum travel distance of thirty feet, or by 20-B extinguishers with a maximum travel distance of fifty feet. The number of extinguishers needed would be determined on placement, provided the maximum travel distance is not exceeded and no voids exist.

OSHA

In the past, the selection and placement of portable fire extinguishers has largely been governed by insurance company requirements and, where applicable, local codes and ordinances. In many cases, these require-

ments were satisfied so long as a few extinguishers
were in evidence at the time of an inspection.

With the passage of OSHA, however, the casual
interest in fire extinguisher protection through volun-
tary standards has become a matter of compliance with
federal law. It is important that mechanical engineers
equip themselves with a basic knowledge of fire protection and portable fire extinguisher installation.

This chapter covers *some* of the legal requirements
contained in NFPA 10, along with general guidance
information on extinguisher selection and placement.

In the case of complex or special-hazard installations, fire protection engineering consultants can
extend the capabilities and services of the mechanical
engineer.

How UL Tests Extinguishers

Underwriters Laboratories, Inc. uses wood crib,
wood panel and excelsior fires to evaluate most
Class A extinguishers. In these tests, the size of
the lumber or other combustibles, the material's
moisture content, ignition and preburn period,
method of attack and extinguishment procedures
are established.

Extinguishment of these fires is primarily
a function of the duration and range of discharge
when using water extinguishing agents. When
other agents are used, the three types of fires
listed below determine the suitability of the
agent and extinguisher. For example, the 2-A
rating of a two-and-one-half-gallon water extinguisher is based on the following three tests:

Wood crib. A crib is constructed of forty-
eight pieces of kiln-dried spruce or fir lumber
having a moisture content of 9 to 13 percent.
The crib consists of 2 $\times$ 2's, 25 5/8 inches
long, placed in thirteen layers of six pieces. After
a total preburn time of ten minutes, the fire
must be completely extinguished.

Vertical panel. A 10-foot $\times$ 10-foot wood panel
is constructed of 1-inch $\times$ 6-inch sheathing, with
3/4-inch $\times$ 3/4-inch firring strips nailed in
horizontal rows spaced 3/4 inches apart. The
panel is ignited with twenty pounds of excelsior
soaked in two gallons of No. 2 fuel oil. The fire
is attacked when the lower firring strips have
burned through and must be completely extinguished.

Excelsior. In this test, twelve pounds of excelsior is arranged in a pile 4-feet $\times$ 8-feet and
1 foot deep. A fuse of n-heptane (two to four
ounces) is used to ignite the 8-foot edge. The
fire is attacked when the flames reach the center
line of the excelsior pile and must be completely
extinguished.

Class B extinguisher tests are conducted on
fires in square steel pans, twelve inches high. For
example, the standard size pan used for a 20-B
listing is fifty square feet. Four inches of water
are poured into the pan, and then two inches of
stove and lamp naphtha is floated on top of the
water, leaving a free board of six inches. (The
stove and lamp naphtha has a specific gravity of
0.718.) The preburn time prior to attack is 60
seconds. The techniques of attack are adapted
to the discharge characteristics of the extinguisher.

To qualify for a Class C rating, an extinguisher
must be capable of being used on energized electrical wiring and equipment without electrical
hazard to the operator. The extinguisher must also
be suitable for use on Class A or B fires or both.
This means that the Class C designation is used
only in conjunction with Class A and/or B ratings.

The test method consists of discharging an extinguisher, which is electrically insulated, through
a 10-inch air gap at a grounding plate that is at a
potential of 100,000 v AC, with no increase in
electrical conductivity.

There are no numeral components for Class D
ratings. The type of combustible metal for which
the extinguisher or agent is applicable and the
area, depth and other characteristics of the fires
that may be controlled or extinguished determine
the best methods and procedures to be used.

The above tests and other evaluation standards
are used in determining the extinguisher ratings
for each manufacturer. Each January, UL publishes these ratings in its Fire Protection Equipment List. The extinguishers are rated according
to class of fire and type of extinguishing agent
for each manufacturer. Copies can be obtained
by writing Underwriters Laboratories, Inc.,
207 E. Ohio St., Chicago 60611.

Sample Problem

A light-occupancy office building needs to be
protected by portable fire extinguishers. The
floor area is 11,100 square feet and of unusual
design (see floor plan below).

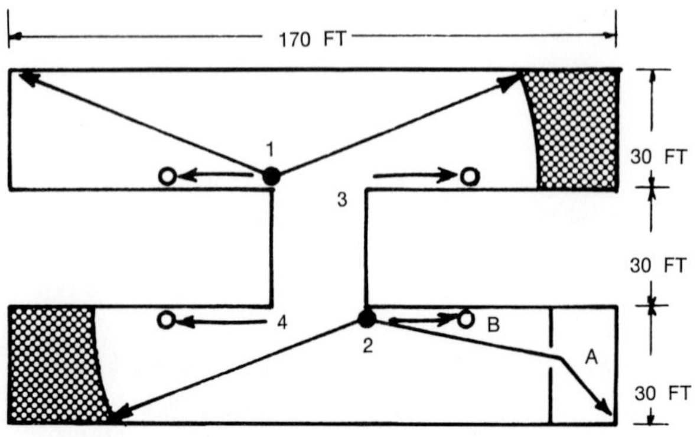

The most common extinguisher selection
would be 2 1/2-gallon stored-water pressure
models rated 2-A. According to Table 17.1, two
extinguishers are needed (11,100 divided by
6,000 = 2). Travel distance requirements are
seventy-five feet maximum.

The two units are placed at Points 1 and 2,
and a check is made on the travel distance requirement. Because of the area's unusual shape,
it is found that the shaded areas exceed the
seventy-five-foot distance. Two additional ex-

tinguishers (at Points 3 and 4) are needed. The
additional extinguishers afford more flexibility
in placement, and alternate locations are indicated. It is important to consider any partitions,
walls or other obstructions in determining the
travel distance.

As an additional item, consider that Area A
contains a small printing and duplicating department that uses flammable liquids. This area
is judged to be an ordinary Class B hazard.
A 10-B:C or 20-B:C extinguisher should be
specified to protect this area.

There are now two alternatives to be considered. First, a fifth extinguisher, either carbon
dioxide or ordinary dry-chemical with a rating
of 10-B:C or 20-B:C could be specified. Second,
the water extinguisher at Point 2 could be replaced with a multipurpose dry-chemical extinguisher that has a rating of at least 2-A:10-B:C.
It should be located near Point B, keeping in mind
the seventy-five foot travel distance for the 2-A
protection and the thirty- or fifty-foot travel
distance required for the Class B protection
that this extinguisher provides.

The final thing to consider involves the nature
of the hazard and the characteristics of the
Class B agent selected. If cleanup is a problem,
then carbon dioxide may be the best choice. If
a corrosion problem could be caused by the agent,
do not use a multipurpose dry-chemical or potassium-chloride dry-chemical.

HVAC Systems

HERBERT WITTE, P.E.

The designing of a building's heating and air conditioning systems is only part of the design engineer's job. He also must incorporate into the systems adequate safeguards against fire, for the source of fire is sometimes heating and air conditioning equipment.

Important factors to consider when providing fire protection for heating systems include where the heat-producing appliances are placed, the combustibility of the structure and its contents, the method of disposing of heat and flue gases, the prevention of excessive pressure or temperature in the appliances, fuel storage and handling, the provision of adequate air for combustion, and a means of preventing fuel explosions within the appliances.

Placement of Heat-Producing Appliances

Boilers, furnaces and other heat-producing appliances should be located so there is adequate clearance between the appliance and the chimney, chimney connector or plenum, and walls and ceiling. This is most important if combustible material is used in the construction of the building. Any material that will ignite and burn is considered combustible, even if it is covered with plaster or given a fire-retardant treatment.

The clearance should be great enough so that the combustible material will not attain a temperature in excess of 194° F. during normal continued operation of the heat-producing appliance. In addition to the obvious fire hazard, continuous exposure of wood to temperatures in this range causes loss of the wood's strength and lowers its resistance to rupture.

Generally, clearances of at least eighteen inches at the top, sides and rear, and from the chimney con

nector, and not less than forty-eight inches at the front (firing side) of fuel-burning appliances are required for furnaces with outlet air temperatures under 250° F. The same clearances apply to low-pressure hot water heating boilers and steam boilers operating at 50 psig or lower.

For other boilers and furnaces the clearances should be at least forty-eight inches above, and thirty-six inches from the chimney connector and all sides and rear. Clearance at the front (firing side) of fuel-burning appliances should be at least ninety-six inches.

A clearance of at least four inches between the appliance and ceilings and walls of noncombustible construction is recommended to avoid possible ignition of combustible material on the other side of the wall or on the floor above. (See NFPA Standards 89A and 211.)

Clearance should be enough to allow adequate ventilation and combustion air, as well as leaving enough room for cleaning heating surfaces and servicing the equipment, including retubing firetube or watertube boilers.

Floor-mounted appliances (with the exception noted in the next paragraph) should be placed either on the ground or on fire-resistive floors with noncombustible flooring or surface finish. No combustible material can be permitted against the underside of the floor or on fire-resistive slabs or arches. This type of floor construction should extend at least a foot beyond the appliance on all sides.

Some factory-built heating appliances have been tested and listed by American Gas Assn. Laboratory and Underwriters Laboratories as safe for installation on combustible floors. They may be installed only in accordance with the conditions of the listing. Appli-

STANDARD INSTALLATION CLEARANCES, INCHES, FOR HEAT PRODUCING APPLIANCES*

Commercial-Industrial Type Low-Heat Appliances — Any and All Physical Sizes Except As Noted		Above Top of Casing or Appliance	From Top and Sides of Warm-Air Bonnet of Plenum	From Front	From Back	From Sides.
Boilers and Water Heaters						
100 cu ft or less Any psi Steam	All Fuels	18	—	48	18	18
50 psi or Less Any Size	All Fuels	18	—	48	18	18
Unit Heaters						
Floor Mounted or Suspended—Any Size	Steam or Hot Water	1	—	—	1	1
Suspended— 100 cu ft or less	Oil or Comb. Gas-Oil	6	—	24	18	18
Suspended— 100 cu ft or less	Gas	6	—	18	18	18
Suspended— Over 100 cu ft	All Fuels	18	—	48	18	18
Floor Mounted Any Size	All Fuels	18	—	48	18	18
Ranges—Restaurant Type						
Floor Mounted	All Fuels	48	—	48	18	18
Other Low-Heat Industrial Appliances						
Floor Mounted or Suspended	All Fuels	18	18	48	18	18
Commercial-Industrial Type Medium-Heat Appliances						
Boilers and Water Heaters						
Over 50 psi Over 100 cu ft	All Fuels	48	—	96	36	36
Other Med.-Heat Industrial Appliances						
All Sizes	All Fuels	48	36	96	36	36
Incinerators						
All Sizes		—	48	—	96	36	36
Industrial Type High-Heat Appliances						
High-Heat Industrial Appliances						
All Sizes	All Fuels	180	—	360	120	120

*Excerpted from NFPA Standard 89M. For information on other types of appliances, and for exceptions to these clearances, consult 89M. Also see NFPA Standard 211 for information on chimney and vent connector clearances.

ances may also be placed on a combustible floor if the floor is protected in accordance with accepted building code practice.

Appliances suspended from the ceiling must have a clearance of at least eighteen inches between the bottom of the appliance and any combustible material.

The design engineer should advise the owner of this hazard, or provide safeguards (such as guard rails) to reduce the hazard. Such appliances must not interfere with the performance of any automatic sprinkler system. For other clearances, consult NFPA Standard 31.

Disposal of Combustion Products

Combustion by-products of fuel-burning appliances should be safely and completely carried outside the building, usually by chimneys. Certain appliances listed by UL and AGA may be connected to a venting system, provided the appliance is approved for use with the venting system.

Chimneys may be either field-erected of masonry or metal, or factory-built. Warm-air furnaces with outlet air temperatures not exceeding 250° F., hot water heating boilers, water heaters and steam boilers operating at not more than 50 psig require chimneys suitable for use with low-heat appliances. Other furnaces and boilers require chimneys for medium-heat appliances.

A low-heat appliance is an industrial appliance such as a commercial cooking range, pressing machine boiler at any pressure, bake oven, candy furnace, stereotype furnace or drying and curing appliance in which materials are heated or melted at temperatures (excluding flue-gas temps) not exceeding 600° F.

A medium-heat appliance is an industrial appliance such as an annealing furnace (glass or metal), charcoal furnace, galvanizing furnace, gas producer, commercial or industrial incinerator, or steam boiler that operates at more than 50 psig pressure when such appliance is larger than 100 cubic feet in size, and other furnaces classified as medium-heat appliances in accordance with nationally recognized good practice. Equipment otherwise classed as medium-heat appliances may be considered as low-heat appliances if not larger than 100 cubic feet in size, excluding any burner equipment and blower compartment.

A chimney or venting system must develop a draft at the appliance not less than that required for safe operation of the connected appliance. If an exhauster is used to increase low draft, automatic shut-off of the fuel-burning apparatus is required in case the exhauster stops.

Chimneys should be supported on properly designed foundations of masonry or concrete. If the load is to be transferred directly to the ground, noncombustible construction having a fire-resistance rating of at least three hours is recommended. (See Figure 18.1.)

Field-erected chimneys must be built in accordance with accepted building code practice. Factory-built

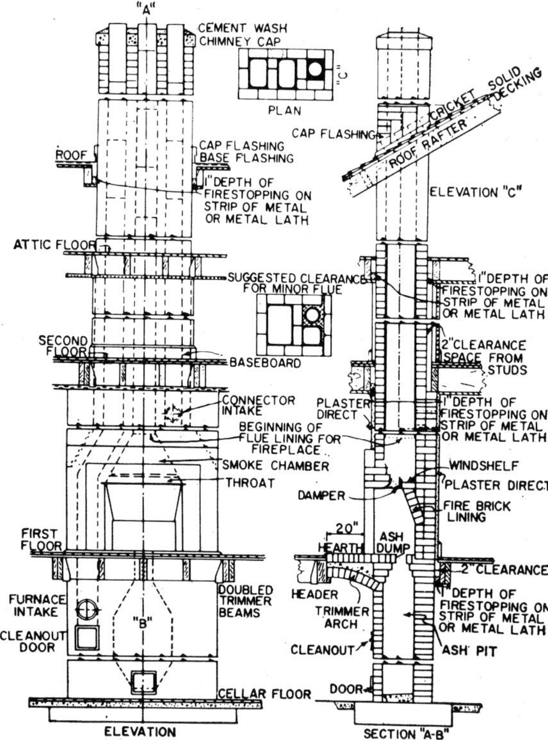

Fire-safe construction of masonry chimneys illustrates good fire protection practices. Note roof penetration.

Figure 18.1.

chimneys and vents should be UL-listed and installed in accordance with the conditions of listing and the manufacturer's instructions.

Combustible material used in the building structure should be at least two inches from outside faces of masonry chimneys. Spaces between chimneys and floors, and ceilings penetrated by chimneys, should be firestopped with noncombustible material. Firestopping of spaces between chimneys and wood joists, beams or headers should be to a depth of one inch only and placed on strips of metal or metal lath laid across the space between the combustible material and the chimney.

If exterior metal chimneys are used with low-heat equipment, the chimneys should have a clearance of at least six inches to walls of combustible construction. At least twenty-four inches are required with medium-heat appliances. The clearance from exterior metal chimneys to walls of noncombustible construction should be at least four inches for chimneys greater than eighteen inches in diameter, and at least two inches for chimneys of eighteen inches diameter or less.

Unless it is insulated or shielded to avoid burning persons who might contact it, no part of an exterior metal chimney should be within twenty-four inches of any walkway, door or window.

An interior metal chimney must be enclosed in a shaft with walls of noncombustible construction in every story above the appliance. These walls must have a fire-resistance rating of not less than one hour. The shaft must provide space on all sides of the chimney for inspection and repair.

Minimum clearance between an interior metal chimney and combustible material in the same story of the building where a low-heat appliance is located must be not less than eighteen inches. At least thirty-six inches of clearance is needed with chimneys for medium-heat appliances.

When a chimney penetrates a roof of combustible material, a ventilating thimble of galvanized sheet steel or other corrosion-resistant metal is usually provided. (See Figure 18.2.) The thimble should extend nine inches above and nine inches below the roof. There should be a clearance of at least six inches on all sides of a chimney for a low-heat appliance and eighteen inches or more for a medium-heat appliance.

A chimney for low-heat appliances only should extend at least three feet above the roof and at least two feet above any portion of the building within ten feet of the chimney. If medium-heat appliances are connected to the chimney, it should be at least ten feet higher than any portion of any building within twenty-five feet.

Chimneys and vent connectors should be short and in a direct line. Sharp turns or other construction features that create resistance to flue-gas flow should be avoided. Chimney connectors should be securely supported to maintain a pitch of at least 1/4 inch per foot of horizontal length of pipe from appliance to chimney.

Vent connectors used to connect gas appliances to Type B gas vents should be installed in full compliance with the terms of the gas vent listing. Substantial galvanized steel, stainless steel pipe with joints fastened with screws or rivets, or refractory masonry is recommended for chimney connectors. Vent connectors of the same piping, such as those used for vertical vents, can provide better vent performance.

It is not safe to pass a chimney or vent connector through a floor or ceiling, because the connector can become damaged or deteriorated within the floor space or in a story above where damage cannot be easily detected. Neither should a connector of any medium-heat appliance pass through a wall or partition of combustible material. (See Figure 18.2.)

Connectors of low-heat appliances may penetrate walls or partitions of combustible material if the point of passage has guards or heat shields. Guards may be metal ventilated thimbles at least twelve inches larger in diameter than the connector to be guarded. Metal or burnt fireclay thimbles built into masonry

TABLE 1 CLEARANCES, INCHES, WITH SPECIFIED FORMS OF PROTECTION*

Type of Protection — Applied to the combustible material unless otherwise specified and covering all surfaces within the distance specified as the required clearance with no protection. Thicknesses are minimum.	Where the required Clearance with no protection is:											
	36 inches			18 inches			12 inches			9 inches	6 inches	
	Above	Sides & Rear	Con-nector	Above	Sides & Rear	Con-nector	Above	Sides & Rear	Con-nector	Above	Sides & Rear	Vent Con-nector
(a) ¼ in. asbestos millboard board spaced out 1″†	30	18	30	15	9	12	9	6	6	3	2	3
(b) 28 gage sheet metal on ¼″ asbestos millboard	24	18	24	12	9	12	9	6	4	3	2	2
(c) 28 gage sheet metal spaced out 1″†	18	12	18	9	6	9	6	4	4	2	2	2
(d) 28 gage sheet metal on ⅛″ asbestos millboard spaced out 1″†	18	12	18	9	6	9	6	4	4	2	2	2
(e) 1½″ asbestos cement covering on heating appliance	18	12	36	9	6	18	6	4	9	2	1	6
(f) ¼″ asbestos millboard on 1″ mineral fiber bats reinforced with wire mesh or equivalent	18	12	18	6	6	6	4	4	4	2	2	2
(g) 22 gage sheet metal on 1″ mineral fiber bats reinforced with wire or equivalent	18	12	12	4	3	3	2	2	2	2	2	2
(h) ¼″ asbestos cement board or ¼″ asbestos millboard	36	36	36	18	18	18	12	12	9	4	4	4
(i) ¼″ cellular asbestos	36	36	36	18	18	18	12	12	9	3	3	3

*Except for the protection described in (e), all clearances shall be measured from the outer surface of the appliance to the combustible material disregarding any intervening protection applied to the combustible material.

†Spaces shall be of noncombustible material.

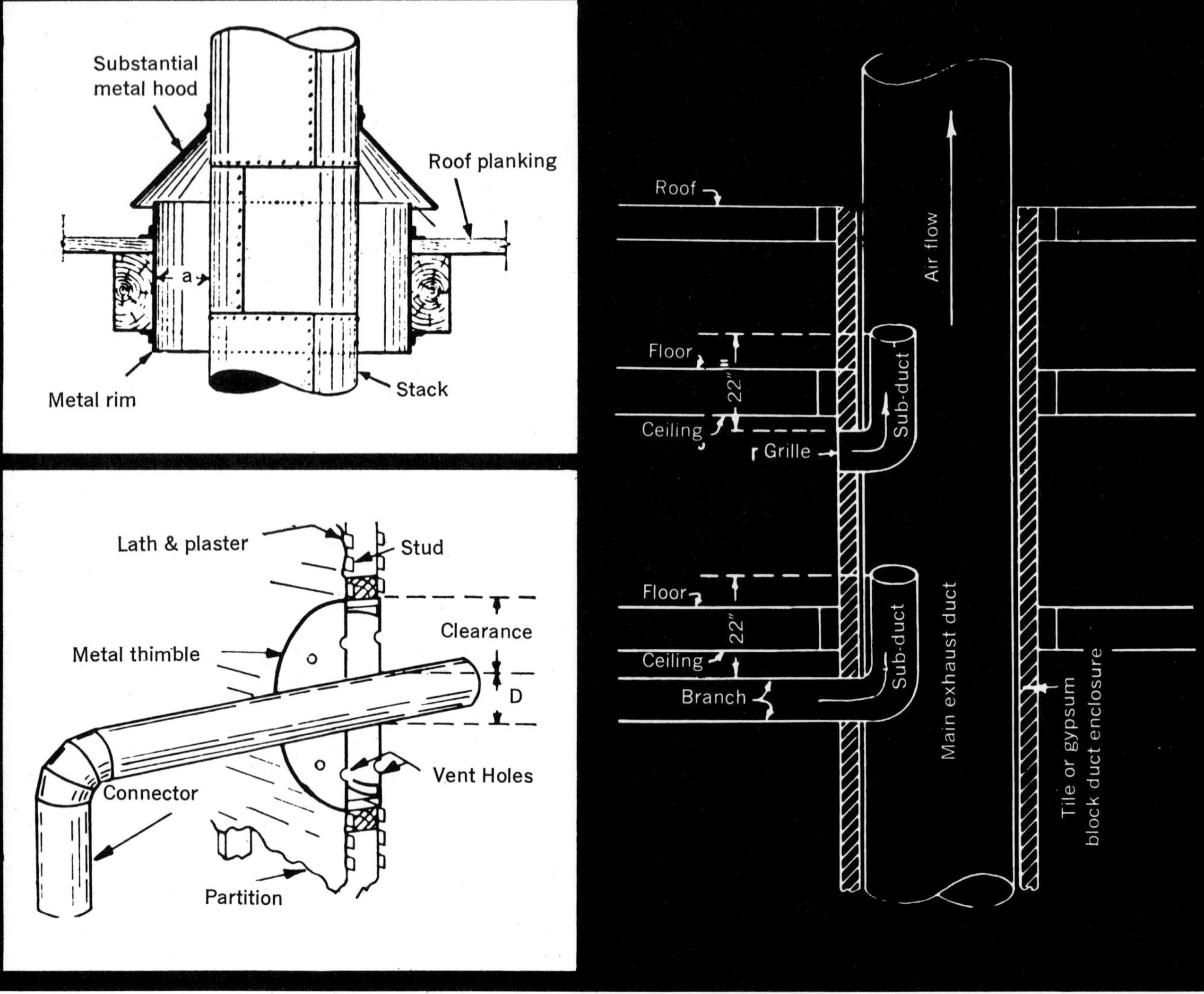

should extend at least eight inches beyond all sides of the thimble.

An alternate method is the removal of all combustible material surrounding the connector to meet clearance requirements: eighteen inches for single-wall metal pipe connectors for low-heat appliances and thirty-six inches for medium-heat appliances. Special piping can be used as connectors with certain appliances. This requires less clearance than for single-wall piping.

Air for Combustion and Ventilation

Adequate ventilation is necessary for the safe performance of heat-producing appliances. For all appliances ventilation is needed to maintain ambient temperatures at safe limits under normal conditions. It is essential that provisions be made to assure adequate air for combustion in fuel-burning appliances at all times.

Preferably, air for combustion and ventilation should be brought from the outside. For example, an air supply inlet in an outside wall of a furnace or boiler room should have a free area of at least ten square inches for each 150,000 Btuh/input of the burners that are firing. If the furnace or boiler room is not adjacent to an outside wall, it is advisable to duct the air from the outside to the appliance room. Fans can deliver ventilation and combustion air, and the air supply should not be blocked. Interlocks should be provided on ventilation and combustion fans to prevent firing of the appliance when the fans are not operating. Appliances should be located to permit free circulation of air in the room. Exhaust fans, kitchen ventilating systems, etc., should not be allowed to deplete the air supply to the appliances.

Fuel Storage and Handling

The engineer should pay close attention to fuel storage facilities, piping for gas and liquid fuels, conveyors for solid fuel, pumps, controls, interlocks and other accessories.

Preferably, tanks for oil fuels should be outside and underground. If outside space is not available, most authorities permit oil tanks to be placed underground beneath a building or in a basement vault. In all loca-

tions, proper venting is required to prevent abnormal pressure during filling and to insure free fuel flow from tanks. Aboveground tanks must also have emergency relief venting to prevent rupture of tanks in case of fire. Design details on tank venting are found in NFPA Standard No. 31.

Containers for liquefied petroleum gas are best located aboveground, even though underground installation is permitted in some localities. An aboveground container can be readily inspected and kept painted to prevent corrosion. Regardless of location, tanks must be equipped with safety relief valves to prevent excessive pressure.

Fuel-handling piping should be substantially supported and protected from physical damage. Corrosion protection is needed for underground piping. Proper allowance should be made for expansion and contraction. Valves, accessories and associated equipment should be carefully designed, sized and coordinated with the fuel-burning apparatus to provide adequate fuel supply at proper pressure and, when heavy oil is used, at the temperature required for good atomization. Interlocks should be provided to prevent attempts to fire if the fuel supply is not in condition for safe burning.

Safety Control Systems

All electric safety controls must be in the safety control circuit when the equipment is operating. To prevent an undetected ground, which could bypass a necessary safety control, the safety control circuit should be a two-wire, one-side-grounded circuit at a voltage not exceeding nominal 120 volts. All switches must be in the ungrounded conductor. In all cases, safety control systems must be tailored to the appliance they serve. For details and examples, see NFPA Standards 85 and 85-B and the UL gases and oils equipment list.

Electrical control circuits should be designed so that operation of a heat-producing appliance is impossible without all required safety controls in the circuit.

All warm-air furnaces, hot-water heating boilers and water heaters must be protected against excessive temperatures. Steam boilers require controls to prevent excessive pressure and should also have a device to protect them against low-water condition resulting

Figure 18.2. Opposite page, top: Table lists information on clearances between heat-producing appliances and walls or partitions with different forms of protection. Center: Typical installation of a factory-made metal chimney passing through a roof of combustible material. Bottom, left: Chimney or vent connectors passing through walls or partitions of combustible material should be installed in the manner shown. Note manner in which metal thimble is installed, and vent holes in the thimble. Bottom, right: Typical arrangement of sub-ducts in a heating system return riser. Dampers may be omitted in such installations; consult NFPA Standard 90A for details.

from prolonged discharge from a relief valve, water leakage from the system, an inoperative water feeder or water line shutoff. A low-water cutoff is the only method to stop firing in the event of low water.

Water heaters and closed hot water boilers should also be equipped with ASME code-rated relief valves having a Btu relief rating not less than the appliance Btu output.

Precautions must be taken to prevent accumulation of an explosive mixture of fuel and air in fuel-burning appliances. These accumulations can be avoided by controls and interlocks coordinated to insure that proper operating sequence is obtained on each start-up, that the fuel is conditioned to permit prompt ignition, and that the proper air/fuel ratio is maintained during lighting off and firing under load. It is essential that there be fuel shutoff valves to prevent the fuel's leaking into the furnace or boiler while the appliance is not being fired.

Air Conditioning Systems

Safeguards against fire must be built into air conditioning systems: ducts can convey fire, hot gases and smoke quickly from one area to another; fire can also enter the building through ducts. Moreover, combustible material used in the duct system can contribute to a fire.

The integrity of the fire resistance of walls, floors and ceilings affected by the duct system must be maintained; the same is true of electrical systems.

Duct System Materials

Materials with relatively slight combustibility and smoke-producing characteristics can create a hazard if ignited under the draft conditions within a duct system. Materials for use in duct systems are classified according to flame-spread and smoke-developed ratings. These ratings are determined by nationally recognized testing laboratories engaged in testing for public safety in accordance with the Method of Test of Surface Burning Characteristics of Building Material.

In general, metal or materials such as clay or cement asbestos is specified for use in ducts. For certain usages, other materials can be used if they meet Underwriters Laboratories, Inc., Standard for Air Ducts, UL 181, and have a flame-spread rating not over 25 without evidence of continued progressive combustion, and a smoke-developed rating no higher than 50. Similar

Opposite page, top: Typical installation of air conditioning system in a building of fire-resistive, protected noncombustible or heavy timber construction. Center: Sheet metal or other protection to reduce required clearance between wall and heating appliance. Bottom: Typical fire damper requirements, showing locations where dampers are or are not required. Right: An arrangement for passing heating ducts through combustible walls or partitions. All diagrams taken from National Fire Protection Association Standards 90A and 90B. For further and more specific information, consult these and other NFPA Standards.

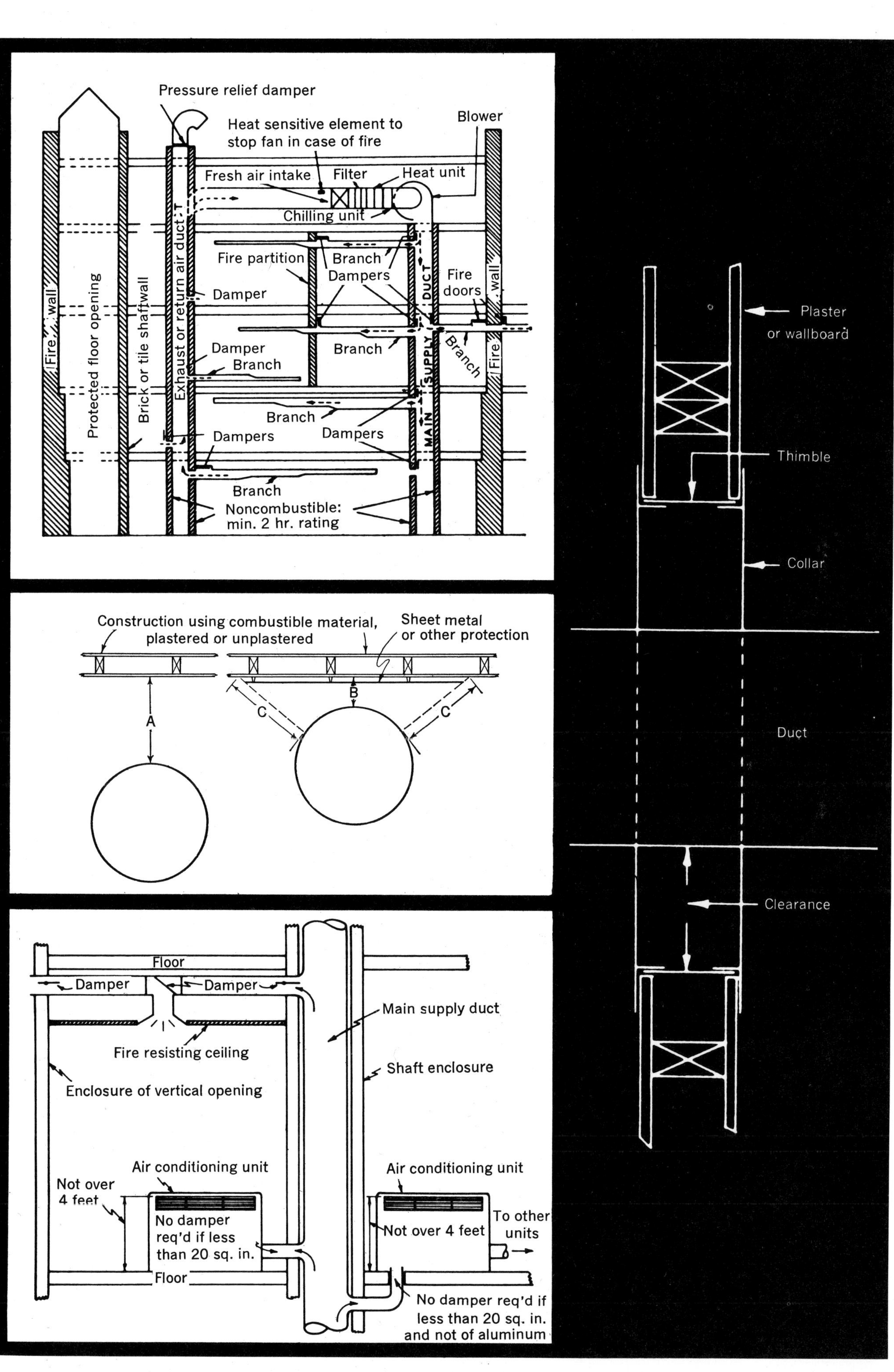

Pressure relief damper
Heat sensitive element to stop fan in case of fire
Blower
Fresh air intake
Filter
Heat unit
Chilling unit
Fire partition
Branch Dampers
Fire doors
Damper
Fire wall
Protected floor opening
Brick or tile shaftwall
Exhaust or return air duct
Damper Branch
Branch
Branch
MAIN SUPPLY DUCT
Fire wall
Branch
Dampers
Dampers
Branch
Branch
Noncombustible: min. 2 hr. rating
Plaster or wallboard
Thimble
Collar
Duct
Clearance
Construction using combustible material, plastered or unplastered
Sheet metal or other protection
A
B
C
C
Floor
Damper
Damper
Main supply duct
Fire resisting ceiling
Shaft enclosure
Enclosure of vertical opening
Air conditioning unit
Air conditioning unit
Not over 4 feet
No damper req'd if less than 20 sq. in.
Not over 4 feet
To other units
Floor
No damper req'd if less than 20 sq. in. and not of aluminum

material may be used for duct connectors. Those not exceeding eight inches in diameter must have a flame-spread rating of not over 50.

To avoid rapid fire spread, duct coverings and linings should have a flame-spread rating of not more than 25 and a smoke-developed rating not over 50.

If a concealed space in an attic, basement or room is used as part of a duct system, it should conform to the requirements specified for ducts. Such space, whether used as part of the duct system or as a plenum, cannot be occupied or used for storage. Fire-resisting ceilings that are part of floor and ceiling construction with at least a one-hour fire-resistance rating may be used as part of a duct system if constructed as tested.

Air filters that do not burn freely or emit a great amount of smoke when attacked by flames should be used. If liquid adhesive coatings are used, high flash-point adhesive should be specified. If flammable liquid is used to flush in-place filters, the system should be designed so that the filters cannot be flushed while the fan is operating. On large systems, an automatic extinguishing system is recommended to protect against fire in combustible material that may accumulate on filters.

Fire Control Through Walls

Where heating and air conditioning ducts pass through walls, floors and ceilings, their fire-resistance integrity must be maintained.

It is best that fire walls (four-hour rating) not be pierced by ducts. If this is unavoidable, automatic fire doors should be installed on both sides of the fire wall at locations pierced by the duct.

With few exceptions, as stated in building codes, fire dampers are required at points where ducts, inlets or outlets penetrate fire partitions. They also are required at each opening in a wall as a fire enclosure of a vertical shaft. If a duct system serves two or more floors, fire dampers or their equivalents are required at each inlet or outlet in the enclosure for a main vertical duct or at each point where the vertical duct pierces a floor it serves.

Dampers should also be installed where each branch duct pierces the enclosure for a main vertical duct. Fresh air intakes to duct systems need fire dampers to prevent entry of outside fires. Fire dampers should be mounted in a substantial sleeve or frame so that disruption of ductwork during a fire will not displace the damper.

Duct openings in a fire-resisting ceiling are to be protected in accordance with the design of the floor and ceiling assembly, as tested by a nationally recognized testing laboratory.

Smoke distributed through ducts can cause injury, death, property damage and panic before fire protection facilities can be actuated by the heat of the fire. Effective smoke detection and controlling equipment is therefore important. The systems should include a means for preventing the spread of smoke by shutting down fans and closing smoke dampers or by actuating an automatic smoke removal system.

Only general principles for fire protection in building systems have been presented in this chapter. For more specific information, the design engineer is urged to make use of the standards published by the National Fire Protection Assn.

References

Code for the Installation of Heat-Producing Appliances, American Insurance Assn., 85 John St., New York City 10038.

National Building Code of Canada, The National Research Council, Ottawa, Canada.

AGA Directory of Certified Appliances And Accessories, American Gas Assn. Laboratories, 1032 E. 62nd St., Cleveland, Ohio.

The following Standards of the National Fire Protection Assn., 470 Atlantic Ave., Boston, Mass. 02110:

Installation of Oil Burning Equipment, No. 31; Installation of Gas Appliances and Piping in Buildings, No. 54; Storage and Use of Liquefied Petroleum Gas, No. 58; Pulverized Fuel Systems, No. 60; Fire Doors and Windows, No. 80; Prevention of Explosions in Fuel Oil—and Natural Gas-Fired Boiler-Furnaces, No. 85; Clearances for Heating Equipment, No. 89M; Air Conditioning and Ventilating Systems, Non-Residential, No. 90A; Chimneys, Fireplaces and Venting Systems, No. 211; Fire Tests of Building Construction and Materials, No. 251; Fire Tests of Door Assemblies, No. 252; Test of Surface Burning Characteristic of Building Materials, No. 255

The following publications of Underwriters Laboratories, Inc., 207 E. Ohio St., Chicago:

Standard for Air Ducts, UL181; Building Materials List; Electrical Appliances and Utilization Equipment Lists; Gas and Oil Equipment List

Industrial Process
Control

WILLIAM H. DOYLE, P.E. and A. J. MERCURIO

When designing for fire-safe operation of equipment that controls an industrial process, attention should be directed to the supplementary devices that take over when the normal operating controls fail. Such supplementary devices can be a major factor in preventing fires and explosions. Process control may also involve properly designing ventilation systems or supplementary protective systems. The basic principles used to solve various industrial process control problems involve the design of controls or equipment to sense and overcome abnormal environmental conditions: heat, pressure, moisture, etc.

Some years ago, there was an explosion in a plant manufacturing an organic insecticide. Part of the manufacturing process mixed chlorine with other material in a kettle. The chlorine reacted with the material as one stage in forming the project. Since this chlorination reaction produced heat, the kettle was cooled by a water jacket and was equipped with a thermocouple that was immersed in oil in a metal tube and inserted into the stirred liquid mass to monitor temperature and permit control of the reaction. If the reaction mass became too hot, the material decomposed.

Corrosion or erosion caused a leak in the thermocouple tube, and the corroded thermocouple incorrectly monitored the process temperature. Eventually the overtemperature situation was recognized. While the process was being shut down and a search made for a new thermocouple tube, the temperature rose, decomposition occurred (more rapidly than was previously thought possible), and the kettle exploded violently.

Six men were killed in the explosion and a seventh died later. The combined property damage and business interruption loss was about $4-1/2 million.

Though this accident was larger and more spectacular than most, it shows that instrumentation, installed only for process and quality control, cannot always assure safe operation too. An inexpensive safety control could have prevented this explosion. Shortly after this accident, many large chemical producers issued internal company standards that required dual instrumentation on processes of this nature.

Industrial Ovens and Furnaces

The fire protection engineer divides industrial ovens and furnaces into three main classes, as defined in the foreword of NFPA Standard No. 86A;

Class A ovens or furnaces are those operating at approximately atmospheric pressures and temperatures not exceeding approximately 1,400° F. where there is an explosion hazard from either, or a combination of, the fuel in use or flammable volatiles from material in the oven or catalytic combustion system; i.e., flammable volatiles from paints and other finishing processes such as dipped or sprayed material, impregnated material, coated fabrics, etc.

Class B ovens or furnaces are those operating at approximately atmospheric pressure and temperatures exceeding approximately 1,400° F.

Class C ovens or furnaces are those in which there

is an explosion hazard due to a flammable special atmosphere being used for treatment of material in process. This type of furnace may use any type of heating system and includes the atmosphere generator. Class C furnaces include vacuum furnaces operating at temperatures from 700° F. to over 5,000° F., at pressures below atmosphere down to 10^{-6} millimeters of mercury, using any type of heating system available.

Although the NFPA standard defines the three classes, its scope is limited to the standards for new installations and alterations of or extensions to existing Class A ovens and furnaces.

Violent, destructive explosions can occur in industrial ovens and furnaces because of malfunctions in the fuel, air, or ignition systems or from ineffective purging, inadequate safety shutoff valving, or flame failure. Devices are available to establish safety control systems that can detect these unsafe conditions and create a fast, safe shutdown of the fuel-burning equipment. A correct installation of a complete safety combustion control system can prevent most of these explosions.

Hazards of Class A Ovens and Furnaces

Firing an unattended steam boiler requires normal controls that monitor the water level and steam pressure. These controls may include other information-gathering devices for stack gas analysis, flow and temperature of fuel, flow and temperature of combustion air, etc. These controls, however, have only incidentally prevented firebox explosions; special instrumentation is needed for that purpose.

The selection of the special instrumentation requires a review of firebox explosion experience to investigate three cause categories: the fuel as it is delivered to the burner; the air system for the firebox and burners; the source of ignition. Faults in any one of these areas can result in an explosion.

It logically follows that the fuel, air, and ignition systems must be given adequate supervisory control. Pertinent pressures, differentials, temperatures, etc., can be supervised by readily available equipment.

Fuel Shutoff

No matter how efficient the various components of the safety supervisory systems, ultimate safety depends on a positive shutoff of the fuel flow to the burners— this is the heart of the safety system.

A fire occurred in a 525-foot-long insulating board dryer. The fire was promptly extinguished by an interior water spray system and the fuel safety shutoff valve was automatically de-energized. After cleanup of minor fire damage, relighting was attempted. When the exhaust fan was started to initiate a seven-minute purge, there was a violent explosion. It severely damaged the dryer and the building and broke a four-inch gas line, releasing burning gas at 30 psig. Seventy-eight roof sprinklers operated to extinguish the building fire. Production stopped for several days, resulting in a loss of nearly $200,000.

The cause: Gummy residue under the seat of the safety shutoff valve prevented the valve from closing completely, allowing gas to fill the dryer ductwork during the cleanup following the initial fire, and the gas found an ignition source when the exhaust fan started. The safety shutoff valve had gone through its proper operating sequence of shutting down when de-energized but failed in its primary function because of the residue under its seat.

The "double block and bleed" or "triple valve arrangement" (see Figure 19.1) minimizes losses from this type of fuel leakage failure, as well as from electrical and mechanical valve failures. This arrangement consists of an approved manual reset or motorized safety shutoff valve installed in the fuel gas line, followed downstream by a second approved safety shutoff valve interconnected electrically with a "reverse acting" 3/4 inch or larger solenoid-operated venting valve installed in a vent line between the valves. This vent line exhausts to the outside atmosphere. During operation of the process heating equipment, the vent valve automatically closes as soon as the circuits are energized so that the upstream safety shutoff valve can be opened.

Purge and Flame Supervision

For safety, the firebox, breeching, and ductwork must be purged of combustible gases and vapors before attempting burner ignition or reignition. Combustibles can accumulate from flame failure, improper ignition, leakage of fuel through shutoff valve, operator errors, etc. Since these accumulations are unpredictable, it is essential that adequate purging be carried out prior to the introduction of an ignition source.

Even with proper control of fuel and air, the burner flame can be lost in several ways—furnace pressure, wrong fuel/air ratios, falling slag, etc. Many of these problems also are unpredictable. Flames can be lost for reasons that cannot be determined by investigations.

If fuel continues to flow into a hot firing chamber
when the burner flame is extinguished, a violent explosion usually results. Consequently, flame sensors
must be capable of distinguishing between flame and
hot refractories and of sensing the small difference
between a stable and unstable flame.

Oven Controls

Drying ovens may use flame-failure instrumentation
when heat is supplied by burning fuel. If the ovens
are used to drive off a flammable liquid, additional
instrumentation is needed to assure adequate ventilation or slow enough input of material being dried so
that an explosive mixture of solvent and air can
never exist in the oven.

The conventional control consists of a flammable
vapor detector that continuously samples the atmosphere in the oven. These devices operate by incinerat-
ing the sample on a hot platinum wire and measuring
the increase in temperature caused by combustion
of the flammable vapor in the sample. A few years ago,
two employees were injured when a switch was pulled
to stop an oven conveyor, and the oven blew up.
Damage amounted to about $75,000. The oven was
equipped with a platinum-wire control designed to
keep the oven temperature below the point at which
sufficient vapors were driven off to bring the flammable vapor concentration to 25 percent of the lower
explosive limit. A different material, however, was
being dried at the time of the accident, and in addition
to flammable vapors, it gave off silicone oil vapor.

This vapor formed an insulating coating on the hot
platinum filament so that its report of a safe oven
atmosphere was erroneous. The only safe way to
specify this type of instrument is to require the manufacturer to guarantee that it will operate properly with
any material to which it may be exposed.

Emergencies Beyond Normal Control

The manufacture of paints requires stirring solvents
into viscous mixtures of resins and oils. If the paint is
not to remain tacky, the solvent must be volatile
enough to evaporate. Normal plant instrumentation
provides some control of solvent addition by metering
as a necessary part of quality control.

An accident at one paint plant proved highly educational. At the plant, miscellaneous flammable solvents
were being pumped from an outdoor tank farm through
a single pipe line to a manifold that distributed solvent
to three separate dispensing stations. Dispensing was
done by gasoline-pump hose and nozzles into process
equipment or other suitable containers. Just before
noon, workers finished loading a container with xylol.
During lunch, the nozzle was left in its triggered position, draining into a pail. The pump at the tank farm
was shut off and the line was draining back into storage.
After lunch, mixed toluol and butyl alcohol were being
pumped when an overflow of solvent onto the floor
was noticed and it was realized that the nozzle left
draining before lunch had not been closed. Fortunately,
the building had good drainage and a good sprinkler
system, and the roof was provided with automatic

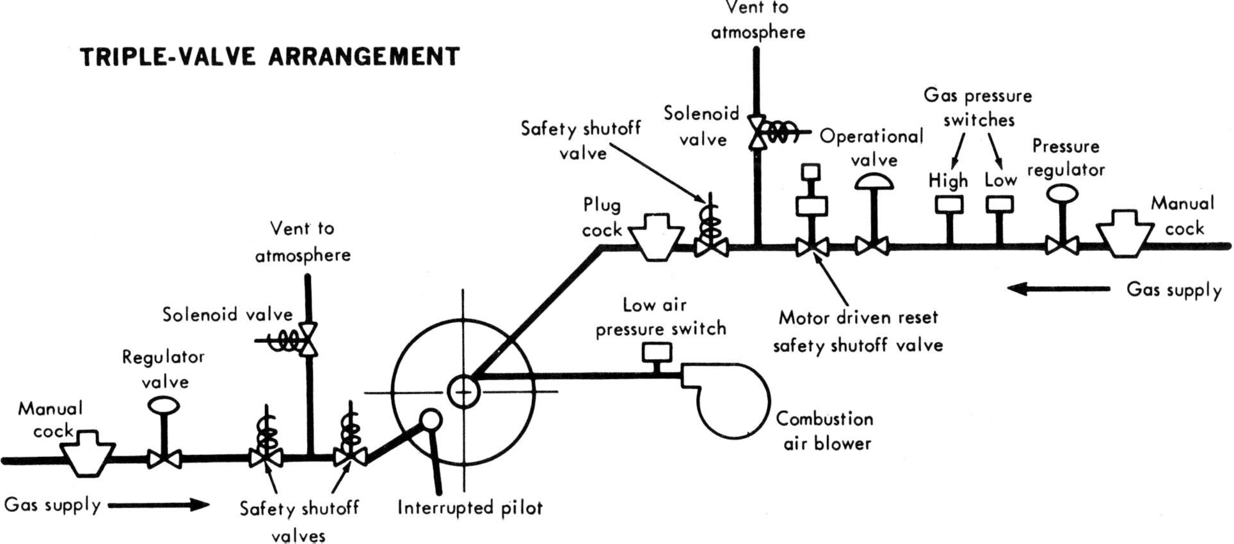

Figure 19.1.

smoke and heat vents. Consequently, when the 100 to 150 gallons of spilled solvent caught fire, the sprinklers operated, washing the fire out of the building; this, combined with the smoke and heat vents, minimized internal damage.

Gasoline-pump nozzles come in two varieties. One can be latched open, and this is what was used at this plant. The type that was initially planned for use was one that cannot be latched open but must be held open by hand for draining. When the purchase order was written, however, the exact kind was not specified; the latch-type was delivered.

Class B ovens and furnaces differ from those in Class A only in that their operating temperatures are higher. The safeguards used for Class B equipment, therefore, are essentially the same. At the recent meeting of the NFPA, a new standard for industrial furnaces (NFPA No. 86-B-T) was adopted. The scope of this standard is limited to Class B industrial furnaces and applies to new installations and alterations of or extensions to existing equipment.

Special-Atmosphere Furnace Hazards

The exacting operating requirements (high temperature and special atmospheres) of various modern heat-treating processes have created more severe operating conditions that call for special furnaces (Class C). These furnaces are hazardous because they are heated by gaseous and liquid fuels and require such combustible special atmospheres as ammonia, hydrogen, endothermic and exothermic gases.

Special-atmosphere, heat-treating furnaces are used to prevent oxidation, to add carbon or to prevent its removal, and for brazing, sintering, bright annealing of copper and steel, and scale-free hardening and annealing of castings.

The special gases used in these furnaces are supplied by one of four methods:

1. *Purchased gas systems* that involve manifolded cylinders or storage tanks.
2. *Exothermic gas generators* in which the fuel gas and air are burned—completely or partially—in a controlled ratio and the products are piped to a furnace as the "atmosphere."
3. *Endothermic gas generators* in which the fuel gas and air—in a set ratio—are passed through a catalyst-filled retort encased in a heating chamber that supplies the necessary temperature to react the gas and air in the presence of the catalyst. The reacted gas is then piped to the furnace.
4. *Ammonia dissociators* (actually an endothermic

generator but, because of its wide use, usually treated separately) in which ammonia vapor, in the presence of a catalyst, is cracked by temperature into its component parts, hydrogen and nitrogen. The dissociated ammonia is then piped to the furnace.

Typical flow diagrams of the last three furnace types are shown in Figures 19.2, 19.3, and 19.4.

The operation and safety controls are generally the same for exothermic and endothermic generators. To assure safe operation of these special-atmosphere furnaces, well-trained operators, versed in all phases of furnace and generator operating, maintenance test and inspection procedures, are needed.

NFPA standards covering Class C furnaces and related generators are contained in NFPA 86C-1974: Industrial Furnaces Using a Special Processing Atmosphere.

Vapor Degreasing Instrumentation Problems

The process theory of vapor degreasing is simple: Cold oily or greasy metal parts are dipped in a layer of hot vapor above a boiling degreasing fluid, the vapor condenses on the metal and the liquid runs off carrying the grease with it. If degreasers use flammable liquids, extensive precautions must be taken against fire. The fluids, however, are usually nonflammable perchloroethylene and almost-nonflammable trichloroethylene. (This discussion is limited to these two fluids.)

Because these fluids are expensive and prolonged exposure to their vapors can cause health problems, degreasers usually have a vented hood and a water-cooled condenser built around the tank near the top. Vapors that do not condense on the work condense on the tank wall and cannot escape. All chlorinated solvents, however, tend to hydrolize on prolonged exposure to water, forming corrosive hydrochloric acid. To avoid water pickup from the air, instruments are used to keep the condenser cooling water temperature higher than the dewpoint of the moisture in the air. The use of these instruments minimizes the moisture condensation and diminishes the corrosion problem but also introduces the possibility of control malfunction. Another special safety control is needed— a thermocouple mounted a few inches above the normal level of the hot vapor to sense the high temperature of a rising vapor level and to shut off the heat to the degreaser.

Another instrumentation problem is illustrated by the following case. Lightning interrupted power to the degreaser area of a manufacturing plant. The power interruption resulted in a loss of cooling water supply. In addition, the solvent boiled off and the heat did not shut off. As the solvent boiled away, the temperature of the remaining fluid in contact with the heating elements increased because the boiling grease and oily residues became concentrated. The degreaser admitted large volumes of black corrosive smoke, and severe corrosion damage to metals occurred over a wide area.

Decomposition reactions of this nature result in fire if there is sufficient oil and grease and insufficient degreasing fluid to keep the mass below its ignition temperature. Boiling of fluid, of course, is one of the best ways to remove heat. In this case, the adage should be revised to "Where there's smoke, there may be fire."

Chlorinated solvents tend to form hydrochloric acid, which reacts with the walls of the degreaser or with the metals being degreased to form metal chlorides. Some of these chlorides, especially aluminum chloride, are catalysts that cause chemicals containing double bonds to react with themselves or with hydrocarbons, such as oil or grease. Since the two degreasing fluids under discussion have a double-bond structure ($Cl_2 C=CCl_2$ or $HClC=CCl_2$), the products of their reaction are a more or less volatile tar, plus extremely corrosive hydrochloric acid.

Since loss of degreasing fluid and consequent overheating result from such things as equipment leakage and removal of solvents as drops on metal parts, a second safety control is needed to shut off heat in case of a temperature rise in the boiling liquid. Controls or process checks of the oils picked up in the solvent bath are equally desirable.

Solvent Extraction

Edible oil recovered from such material as cotton seed and soybeans was once squeezed out by mechanical pressure. This process, however, left considerable oil in the pressed cake and now has been supplemented with or supplanted by treatment of oil-bearing seeds with a solvent that dissolves out the oil.

EXOTHERMIC GAS GENERATOR—TYPICAL FLOW DIAGRAM

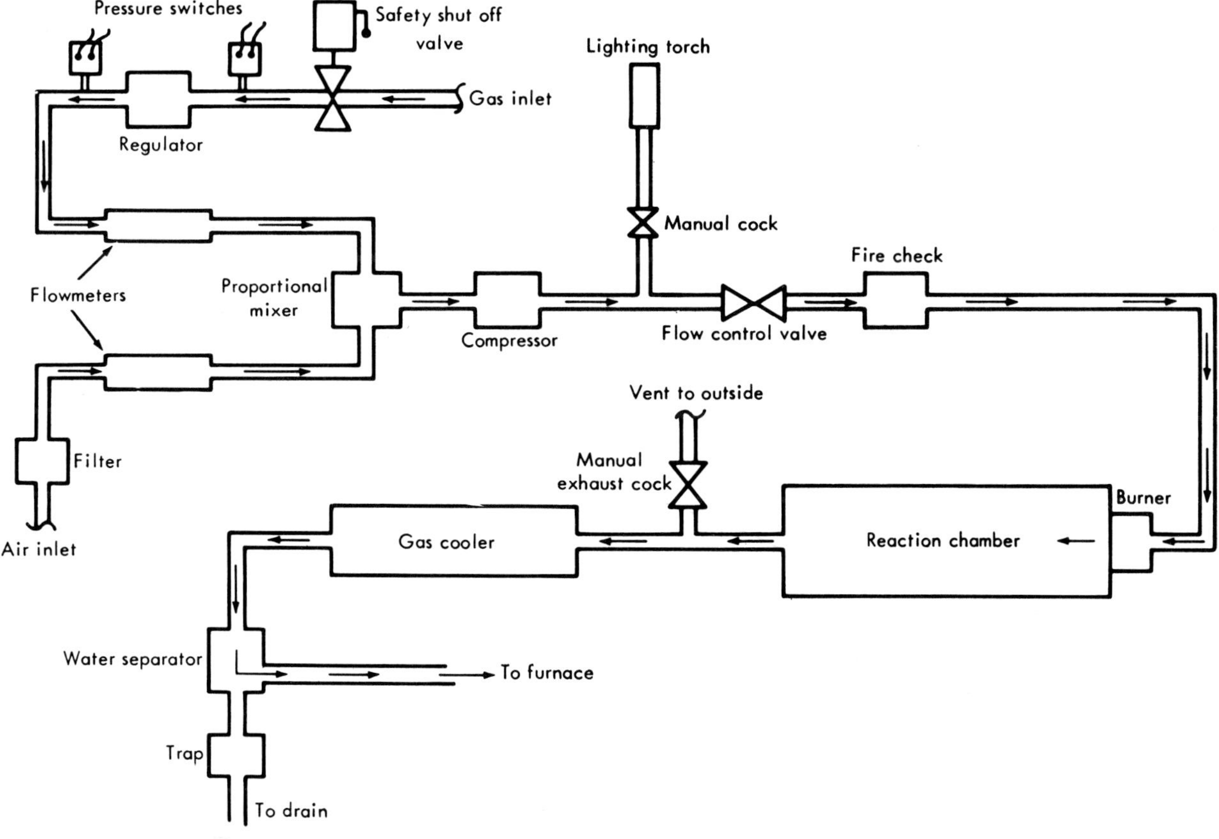

Figure 19.2.

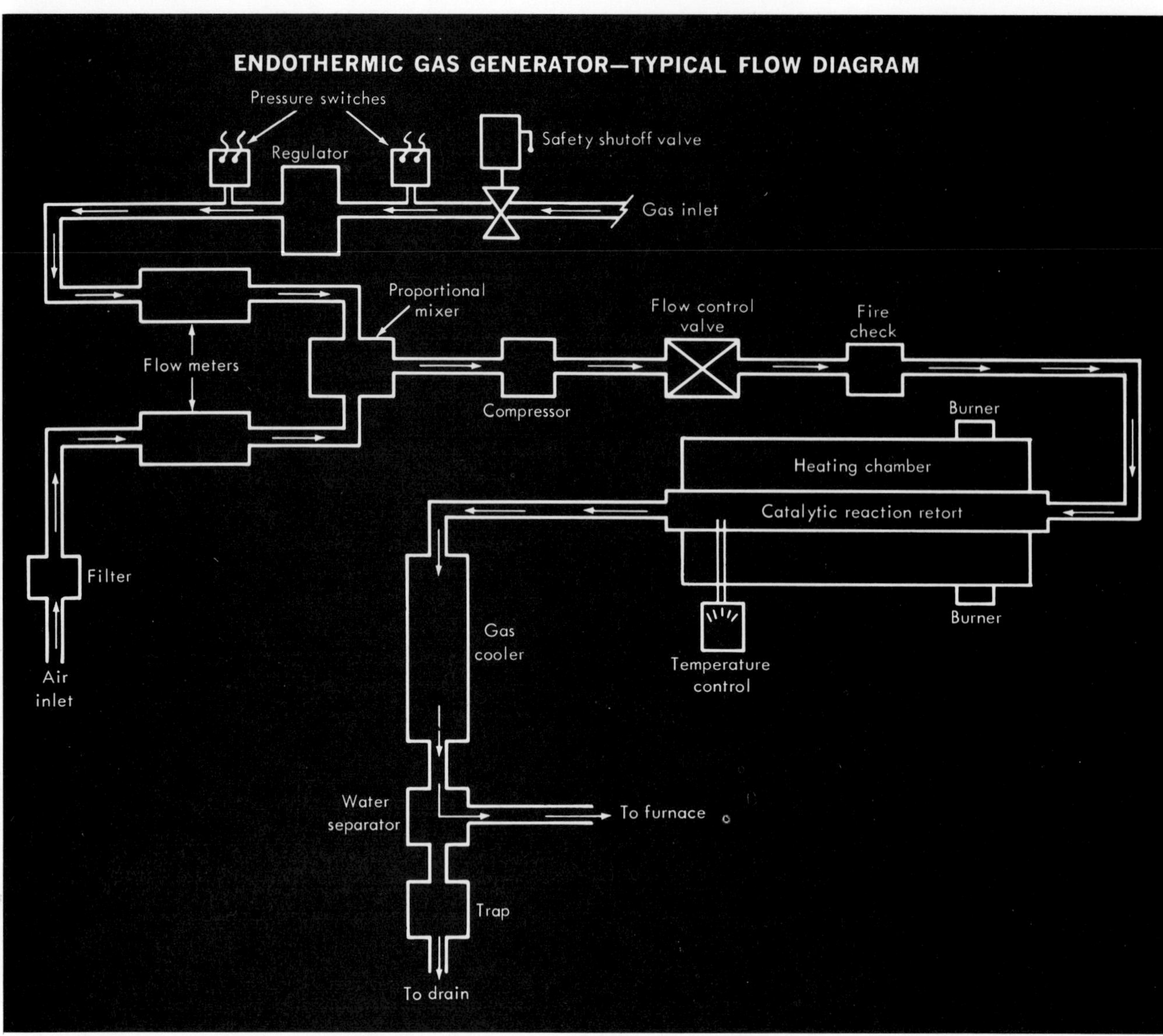

Figure 19.3.

Attempts to do this using nonflammable chlorinated solvents, while technically successful, have proved impractical. The cake from which the oil is extracted is largely used for cattle feed and, if all the chlorinated solvent is extracted, very long and uneconomical drying times are required. On the other hand, if economically unimportant traces of chlorinated solvents are left in the cake, prolonged feeding will cause the cattle to develop chlorosis. Accordingly, the solvent normally used is hexane, since traces of hexane in the cake are not toxic. Certain controls are needed to assure that flammable hexane stays within the equipment at all times and that prompt warning is given of its release or of conditions that might lead to its release.

Once hexane is in the extraction machinery for a short time, all vapor space inside the equipment becomes too rich to burn. In the process of becoming too rich, however, the vapor space passes through the explosive range. An oxygen analyzer should be provided and inert gas fed into the equipment before hexane is introduced. The flow of inert gas should continue until the oxygen analyzer shows that the oxygen content of the gas being purged out of the equipment is less than 10 percent. The oil seed generally is fed into the equipment through a screw conveyor that has one or two flights cut away so that seeds solidly choke part of it and act as a plug against vapor escape. Interruption of seed flow to the screw, however, can result in loss of the plug. A flammable vapor detector should be provided at this point to indicate vapor leakage.

As the de-oiled seeds leave the equipment, they

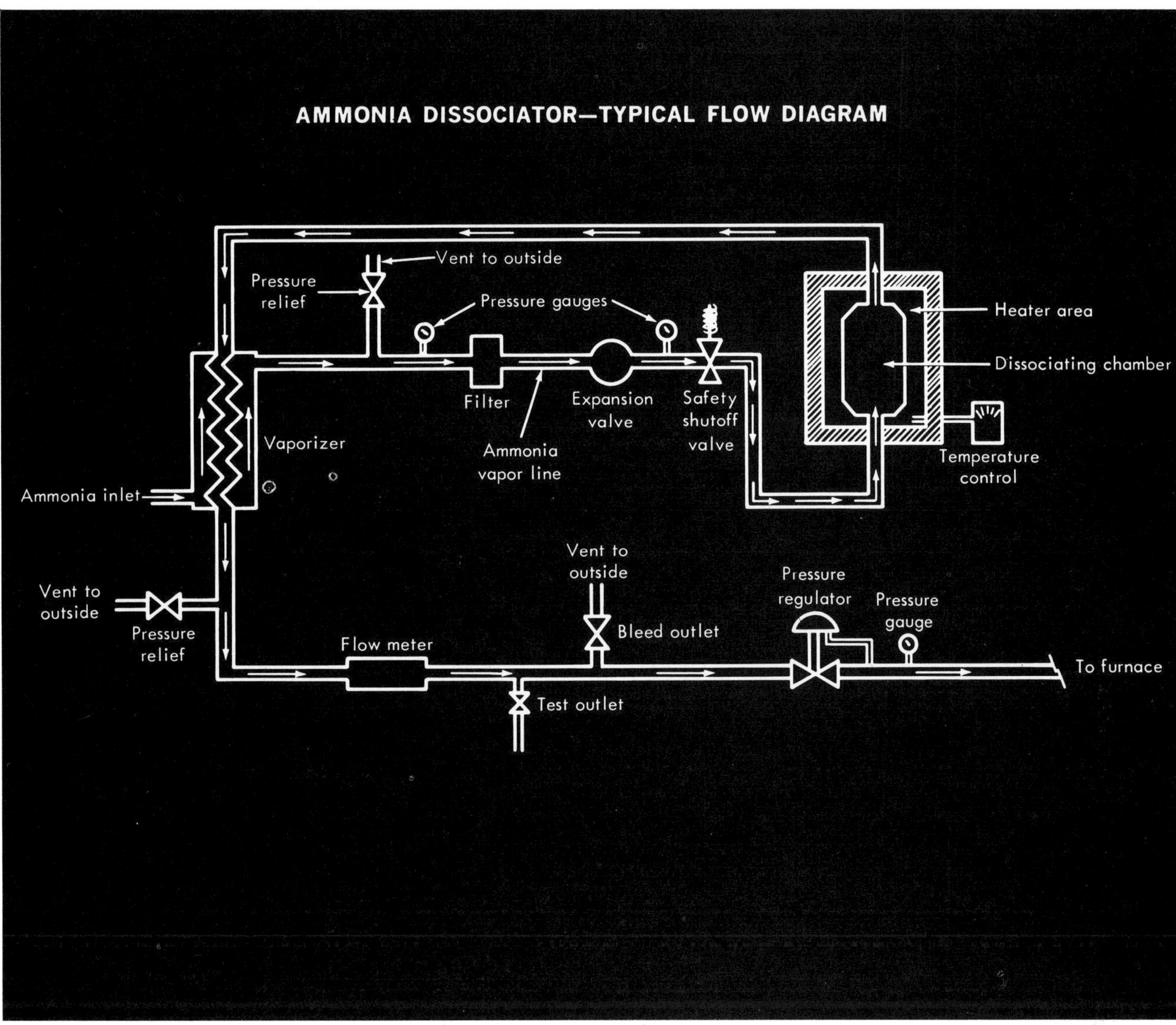

Figure 19.4.

must pass through a heated dryer to vaporize the solvent. This drying operation is also conducted in an atmosphere too rich to burn, and the hopefully solvent-free seeds leave through another choked conveyor, which is another point where a flammable vapor detector is needed. The vapors from the drying section pass to a water-cooled condenser that liquifies the hexane and returns it to storage. Cooling water failure causes uncondensed hexane vapor to escape here, a circumstance that necessitates another vapor detector and also a flow indicator or high-temperature sensor to indicate loss of cooling water.

Oil is removed from the hexane by boiling off the hexane in a still. The solvent condenser on the still requires the same treatment as the one on the seed dryer. When the process is shut down for repairs, it should be thoroughly purged to remove all the traces of solvent vapor before any equipment is opened. For this purpose, portable flammable vapor indicators are used.

There are several points where automatic control should be provided to detect the presence of flammable vapors. Vapor release should be checked in the vicinity of such other pieces of equipment as solvent pumps, manholes, valves, etc. One type of solvent detection equipment makes use of metallic tubing and a pumping system that draws a sample from the point being tested to the analyzing instrument. By appropriate valving, up to twelve sampling tubes may be supervised by one analyzer. Units of this type, though, are usually set for a thirty-second sampling time for each sampling tube, which means if twelve points are checked, there will be a six-minute lapse between samples from any one point and six minutes can be as bad as six days if a large spill occurs. Accordingly, it is better to use a diffusion-head unit in which a small on-the-spot analyzer detects the vapor as soon as it is admitted and electronically reports the location of the hazard back to a central panel.

Control equipment installed in areas that may be hazardous because of the presence of flammable gases, vapors or dusts presents another problem. Pneumatic instrumentation systems are now often regarded as old-fashioned, but they are quite suitable for such applications.

The use of fluidics similarly permits hazard-free installations. Electrical and electronic instrumentation, on the other hand, may introduce hazards. Where this is a problem, the NFPA Standard No. 496, Purged and Ventilated Enclosures for Electrical Equipment in Hazardous Locations, and No. 493, Intrinsically Safe Process Control Equipment for Use in Hazardous Locations, provide guidelines. Various publications of the Instrument Society of America provide background information and more detailed guidelines. Fortunately, however, in the majority of cases it is not necessary to consider this particular facet of industrial control.

Key Word: Prevention

The fire protection engineer is interested in process controls when he puts on his alternate hat and thinks of explosion prevention. Chemical reactors must be kept from exploding, fuels must be burned under conditions controlled by instruments that are quick to detect trouble and quick to take remedial action. Properly designed and installed instruments will work better than the best human operator. Finally, instru-ments must prevent the unintentional mixing of flammable liquids or gases with air. If this prevention is impossible, instruments must alert human operators to a developing danger—even if only to warn them to beat a hasty retreat from the danger area.

References

Handbook of Industrial Loss Prevention, Chapter 37, McGraw-Hill, N.Y.

Fire Protecton Handbook, Chapter 9, NFPA, Boston, Mass.

NFPA Standard 86-A, "Ovens and Furnaces, Design, Location and Equipment," *National Fire Codes,* Vol. 9, NFPA, Boston, Mass.

NFPA Standard 86-B-T, *Industrial Furnaces 1968 NFPA Technical Committee Reports,* Vol. 1, NFPA, Boston, Mass.

Recommended Good Practice for Safeguarding Class B and Class C Furnaces and Ovens, FIA, Hartford, Conn.

Metal Degreasing with Chlorinated Solvents, DuPont, Wilmington, Delaware.

R. F. Schwab, *Trichloroethylene & Perchlorethylene Degreasing,* Bulletin N-36, FIA, Hartford, Conn.

NFPA Standard 36, "Solvent Extraction Plants," *National Fire Codes,* Vol. 1, NFPA, Boston, Mass.

Occupancy Fire Record, Oil Extraction Plants, NFPA, Boston, Mass.

E. P. Cofield, Jr., "Solvent Extraction of Oilseed," *Chemical Engineering,* Jan. 1951, pp. 127–140.

O. R. Sweeney, L. K. Arnold, E. G. Hallowell, *Extraction of Soybean Oil by Trichloroethylene,* 1949, Iowa State College Bulletin, Ames, Iowa.

Ernest C. Magison, *Electrical Instruments in Hazardous Locations,* 1966, Plenum Press, N.Y.

Part IV

Fire Alarm Systems AND Other Electrical Fire Protection Problems

Fire Alarm Systems

PATRICK E. PHILLIPS, P.E.

Where there's smoke, there's fire, but usually there's also someone too excited to call the fire department. Only rarely are fire detection systems installed around true engineering concepts to minimize loss of life or property. Occasionally they are installed for fire insurance purposes. But most systems are installed only because building codes require them.

Building codes and standards cite the minimums required, but these same codes and standards are almost invariably cited as the maximum desired in the design. The design engineer consequently finds himself in an uncomfortable position where his minimums are his maximums. Because code-and-standard-required systems rarely provide the best design for a specific building, how can the engineer determine how to design and specify a good fire detection system?

Manufacturers' brochures often elaborate on how well their detector protects a building. This is inaccurate because the terms "detect" and "protect" are not synonymous. If the engineer keeps that in mind, a superior system design will be relatively simple.

All fire detection systems have one basic purpose—to get a person or system to do something about the fire. After detection, the system must transmit an alarm. Preferably the alarm initiates an action that puts out the fire. Without the full cycle—detection, alarm, extinguishment—the system is of limited value.

A fire protection signaling system must perform four significant functions to accomplish its intended purpose. First, the system must detect a fire rapidly, before there is significant damage. Second, it must start a sequence of events to intelligently evacuate occupants of the building who might be endangered by the fire. Third, it must transmit an alarm or signal to notify some responsible party (or an automatic

system) to start extinguishment. Fourth, it should check itself for operability.

Manual Devices

Where manual stations are the only means of transmitting an alarm, standards require that manual alarm stations be unobstructed, readily accessible, in the normal path of exit travel and not more than 200 feet apart. In buildings where the path of exit might be long and obstructed by cross corridors, it is advisable to install manual stations at lesser intervals.

Manual systems may also incorporate watchman supervisory service. In this case, the watchmen must visit each station during their tours of the building. The standards do not give detailed guidance on the location of these stations, but to be effective, the watchmen must be able to see all portions of a building, including inside every room. Although insurance authorities generally require watchmen to tour the entire building every hour, more frequent tours are desirable. Because the costs of watchmen are generally too high to justify, this service is rarely used. But, if watchmen are used, they should be given the exclusive responsibility of surveillance and should not have other duties, such as night-receiving dock operator, janitor, or repairman.

Except when required by local codes, there is no need to superimpose a complete manual alarm system throughout a building that has a complete automatic alarm system. If the system is also an extinguishing system, the need for manual stations is further reduced. Few totally sprinklered buildings have manual fire alarm stations. A rule of thumb for buildings with automatic systems would be one box at each major exit on each floor, usually two stations on each floor. For large buildings, more stations would be desirable.

Other factors to consider are the type of construction, number of exits, and "value density," a term to describe the dollar value per square foot plus intangible and often uninsurable value (e.g., programmatic importance, loss of prestige and customers who go elsewhere). Where value density is high, more protective measures are justified.

Automatic Detection Monitoring

To provide continuous surveillance, automatic detection devices may be incorporated into any alarm system. The devices may be sensitive to any reasonable change in environment.

Automatic detection may include devices activated by extinguishing systems. The water-flow switch of a sprinkler system is a detector, but so is the pressure switch connected to the manifold of a gas-operated extinguishing system. Other devices that determine the system condition or readiness are also detectors; they are normally called "system supervisory devices" and can be found, with water-flow switches, in Underwriters Laboratories, Inc.'s Fire Protection Equipment List under the heading "Extinguishing System Attachments."

Alarms

After the detection system has found a fire, it must alert someone to do something about it. Audible alarms are used to alert occupants who may respond by starting an extinguishing process or by evacuating the building or fire zone.

Audible general alarms are usually avoided in situations where panic may result. These include hospitals, auditoriums, and other large assembly buildings. In such cases, the alarm is not generally sounded but is sent automatically to supervisory personnel and to the fire department.

NFPA Standard 72A for local fire alarm systems defines a local system as "one which produces a signal at the premises protected." While alarm-sounding devices are required for fire alarm systems, the Standard does not control the placement or the audibility level of such devices. The specification writer should assure that these items are covered to satisfy individual project needs.

Preferably, the signal should be transmitted to a point where action can be taken. This could be to a local watchman who could call the fire department or, in a very large plant, a "proprietary" office, where men are on duty at all times to receive these alarms and take action.

Locating the Fire Zone

If the premises are large and complicated and if trained personnel are available, coded local alarms are valuable to help locate the fire. Coded signals will direct firefighting personnel to the fire more quickly. Because coded systems are more expensive, the engineer should closely analyze the need for fire zones. Special industrial processes, high value-density areas, and high life-loss areas should be of special concern and may require separate annunciation. Multiple-story buildings usually need annunciation for each floor.

Most coded alarm signals, within a building, are single-stroke bells, but they could be chimes, klaxons or even voice tapes. Because alarms serve not only to direct action to the fire but also to evacuate people, they should be "alarming." A soft, slow-stroke bell system usually does not shake people. Voice transmission of specific directions should be considered as an improvement over simple bells or alarm signals.

Coded signals can be transmitted to outside agencies such as fire departments. In some systems, transmitting a coded signal is mandatory in order to differentiate a signal coming from separate buildings or areas. But transmitting a coded signal is useful only if it tells the recipient something that is necessary to minimize loss of life or property.

An alternative to transmitting coded signals (or a way to elicit quicker response) is to provide an annunciation panel located at the premises. This panel shows the fire department what area or zone is on fire or in danger. The panel should be prominently situated, preferably outside the building at the main entrance.

The number of coded signals or annunciation points can be reduced in buildings with automatic extinguishing systems. Multiple zones are useful if fire fighters must be directed to the area to reduce the loss. If the extinguishing system is properly designed, it should control the fire and reduce the need to hurry to the scene.

For a sprinklered building with no special annunciation needs, one fire zone per floor per riser might be adequate, but other features should be considered to determine zoning needs. Here are some questions that should be answered to come to a proper determination:

1. Are automatic extinguishing systems used in all areas?
2. Do we have special detection systems supplementing the protection system?

3. Are there areas of detection only?
4. Is there a need to annunciate separate occupancies in a multiple-occupancy building?
5. Is there a particularly hazardous area that will prevent immediate re-entry? (This could be an explosives area or one where toxic materials are handled.)
6. Is there a specially critical area that must be manned continuously or at least will require immediate re-entry?

Signal Transmission

Alarms may be transmitted to a central station (NFPA 71), to a proprietary station (NFPA 72D), or to the fire department's municipal alarm system (auxiliary system—NFPA 72B), or to the fire department via a special circuit (remote station system—NFPA 72C).*

Usually, a central station is considered the most desirable, but the insurance rating authority should be consulted to verify that the central station office is accredited. The term central station means signals are received at an approved central station office where two or more experienced operators interpret signals and take necessary action. This includes not only contacting the fire department by telephone but also sending a "runner," usually an armed guard, to the protected property to minimize loss.

A central station must be an independently owned firm whose principal business is to provide and maintain a protective signaling service. It must transmit all alarms over supervised circuits arranged so that a single break or ground fault will not cause a false alarm and will not prevent transmission of a true alarm. These are usually wired as a "McCulloh circuit," a circuit that normally transmits over the two wires but can be switched, manually or automatically, to transmit over one wire and ground. With this capability, a break or ground fault of a single wire will not render the system inoperative. This concept may be nullified to some extent by a carrier-leasing agency providing other than individual wired circuits.

If an auxiliary circuit is used, the building can be designed with common local energy systems or with a shunt-loop system. Shunt-loop systems are usually frowned upon because they require running the municipal alarm system circuit into the building. The current standard limits the length of municipal circuit in private premises to 750 feet. When the NFPA member-

ship last considered this standard, there were strong feelings that this limit should be 3 feet.

Shunt systems must always be installed in conduit or electrical metallic tubing while local energy systems (or any other standard system designs that do not jeopardize the municipal systems) may take advantage of special "limited energy circuit cable" rules that permit installation of special wire under specified conditions without conduit.

Many designers select this type of cable because it saves conduit costs and increases system flexibility. In general, its use is permitted if "not less than 7 feet off the floor and if adequately protected against injury." This does not preclude its being left exposed after installation: the "adequate protection" can be open air if it is unreasonable to expect traffic or operations that will damage the cable. Manufacturers of this cable are shown in UL's Electrical Construction Materials list under the heading "Wires, Miscellaneous (Guide No. 460 I 90)."

Remote-station protective signaling systems are the most recent addition to the fire alarm field. These systems were not recognized in the NFPA Standards until June, 1954. The system is connected to a receiving point remote from the protected property, usually by a leased telephone pair. The leased circuit is electrically supervised so that a single open or single ground fault or loss of electrical power will transmit a trouble signal, not a false alarm. Because a McCulloh circuit is not required, an open circuit will prevent transmission of a fire alarm.

Remote-station alarms are usually transmitted to the fire department, but the authority having jurisdiction may accept alternative locations—such as a police department or even a doctor's telephone answering service, taxi cab office and funeral parlor where someone is on duty twenty-four hours a day. These alternate locations must have a direct means of contacting the fire department and usually are the location called by the general public.

Although the remote-station system is often described as "direct connect," this is usually not accurate. Because it most often uses leased telephone wires, all circuits must funnel back to the telephone exchange before reaching the fire department.

Some large industrial plants have what are called proprietary protective signaling systems, but the name is often improperly used, especially by plants that transmit alarms over a local alarm system to their own fire departments. Because a plant owns its own equipment does not make the system "proprietary."

If remote stations or auxiliary systems are to be designed, the fire department should be contacted to verify that it is willing to accept the signals and

---

*Current NFPA standards are written primarily to describe alarm transmission. The words in each standard differ, with minor exceptions, in how and where the alarm is sounded but are quite similar in wording for detector location and wiring within the building.

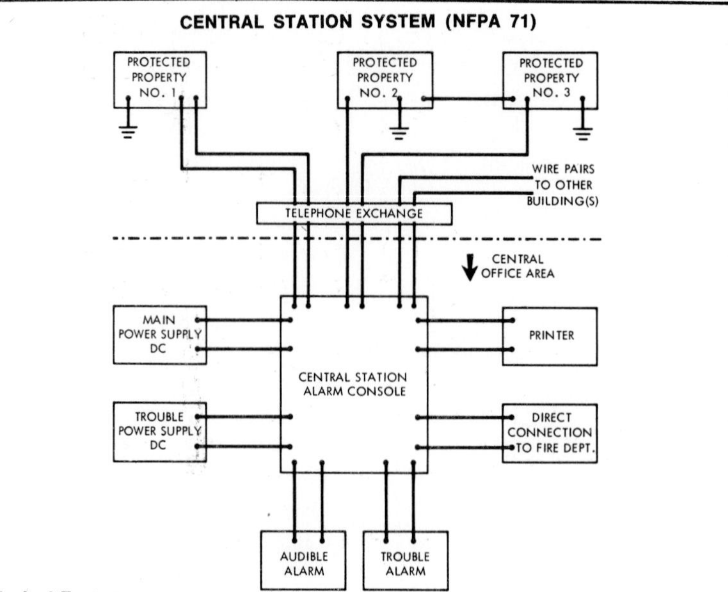

Typical Features:

Leased equipment, no capital investment.

System design, installation, maintenance and test by the central station.

If an Underwriters Laboratories-listed central station office—insurance credits are possible.

Recorded coded signals, experienced observer-operators.

Essential transmission circuits arranged so that a single break or ground fault does not cause a false alarm and will not prevent transmission of a true alarm (McCulloh circuit or other alternate path).

Runner, usually armed, sent immediately to take action to protect occupant property.

It can detect and alarm for other, nonfire conditions—burglar, building temperature, water pressure, loss of electrical power, condition of automatic extinguishing system features.

Direct-line voice communication with the proper fire department to explain in detail the fire or other emergency. This is usually done by direct-wired telephone, with a requirement for an alternative means of contacting the fire department.

Coded signals are usually "noninterfering."

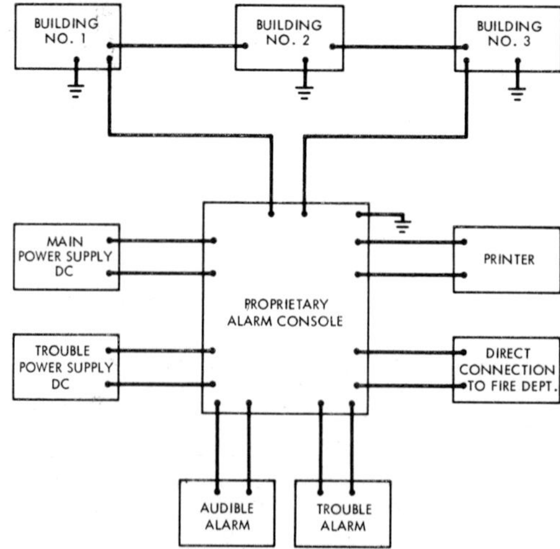

Typical Features:

It is practical only for large facilities because it requires constant attendance by two trained men, plus runners. Normally a "Class A" circuit is required.

All alarm signals must be recorded at the alarm office, preferably automatically.

Alarm office should be in a separate detached fire-resistive building.

Two means of alarm transmission are usually required, the one being a direct supervised line and the other either a municipal fire alarm box within 50 ft or a telephone that does not go through the plant switchboard.

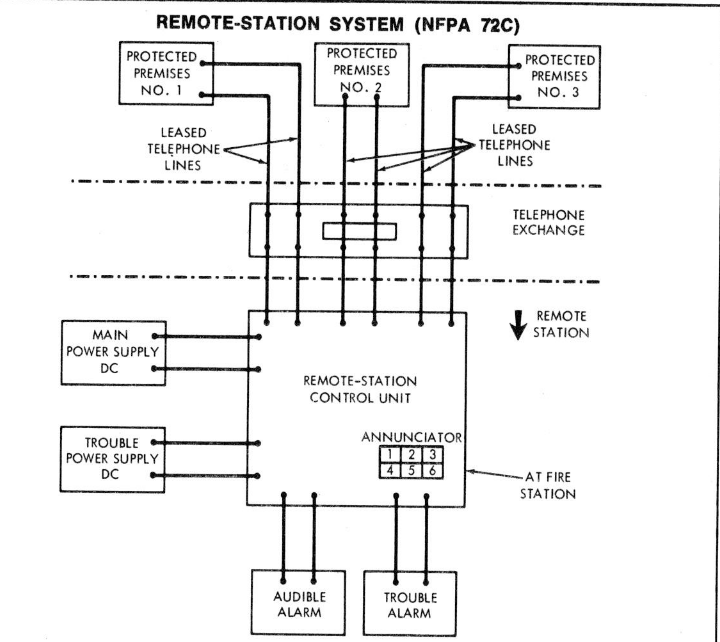

Typical Features:

Usually a direct connection (uncoded) between the transmitter and the receiver.

Equipment may be bought or leased. Leased systems usually incorporate design, installation, testing and maintenance.

The fire department may be unwilling or unable to receive trouble or supervisory alarms. This will require a separate transmitter, transmission lines and receiver.

Multiple-alarm transmissions require separate circuits or coded signals.

Number of automatic detection devices in a single circuit should not exceed 100.

AUXILIARY SYSTEM/LOCAL ENERGY (NFPA 72B)

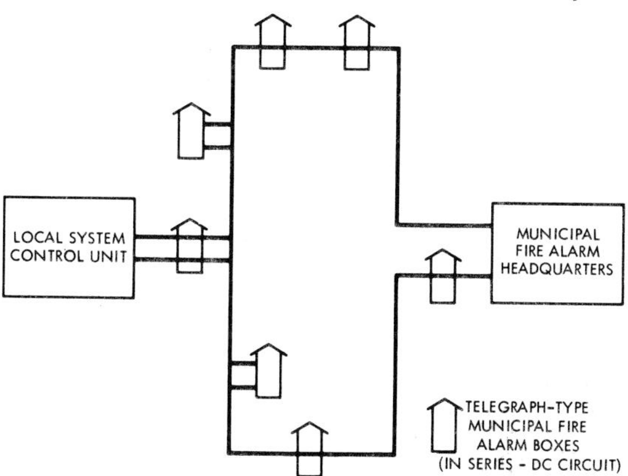

Typical Features:

System alarm transmission is only as good as the city auxiliary loop.

Alarm transmission is a coded signal.

Equipment in the building is usually the property of the building owner, but the coded transmitter, though bought by the building owner, usually becomes the property of the city.

The city is usually responsible for maintaining the loop up to and including the transmitter. Interior maintenance is the owner's responsibility.

The interface of these responsibilities is sometimes a problem.

Auxiliary loop connections may be unavailable or reserved for city property and schools.

Maintenance is often a problem. To be considered standard, a system should have periodic maintenance, inspection and tests. If the system is installed for insurance credits, most rating bureaus require that this be a contract for monthly maintenance by an authorized installer. Auxiliary loops are usually maintained by city personnel and, as a consequence, may not be acceptable for insurance credits. The arrangement and requirement of the city auxiliary loop are shown in NFPA Standard 73. .

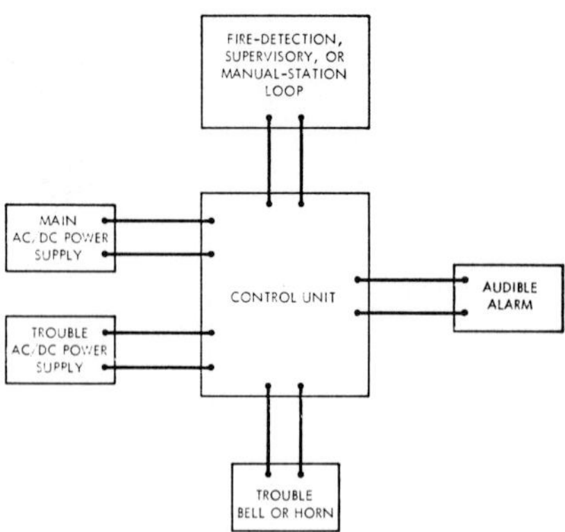

Typical Features:

No alarm transmission to outside agency.

Internal system only.

Minimal insurance credit, if any.

Costs within the building are almost the same as other systems that transmit alarms. Additional costs are primarily for a transmitter and transmission lines and receiver, if they are necessary.

All equipment is usually purchased.

Outside maintenance and test contracts are needed to make sure the equipment is in good operating condition. This service usually costs as much as a system that can transmit an alarm to an outside agency.

The system is ineffective when the building is vacant or unattended.

that it can provide an acceptable response. Many big-city fire departments refuse signals from other than an accredited central station because the alarm offices are already overloaded, and auxiliary system connections may be reserved for municipal buildings, schools and places of assembly. In some cities, the municipal system may be inadequate or in poor repair.

A proprietary system is like a central station office but is reserved for one large customer. Almost all requirements for a central station apply to the proprietary office, including two operators on duty at all times. For a Class A system, the McCulloh circuit features are required, but the Class B circuit does not require this capability and, therefore, is considered the lower ranking of the two. As the proprietary office is only the receiving station, the fire department must still be summoned. This must be by a positive means— usually by direct connection over wires electrically supervised and under the control of personnel at the protected premises. In some instances, a municipal fire alarm box within fifty feet of the proprietary office may be acceptable. A backup by ordinary telephone independent of plant switchboard is recommended.

Fire Detection Systems

PATRICK E. PHILLIPS, P.E.

Automatic detection devices are numerous in type, shape, size, and price. They may be simple fusible material arrangements or complicated electronic devices. Although they all are fire detectors, they are triggered by different elements of a fire: flame, convected heat, radiant heat, smoke, or a combination of these.

Fire detection systems are sometimes justified because they operate more quickly than sprinklers and will sense the fire early enough so that it can be easily extinguished manually. A review of the test methods used reveals that all detectors at maximum spacing will have the same sensitivity as a sprinkler head over a large fire.

Thermostats

UL Standard No. 521 describes how fire detection thermostats are tested. (See page 171.) The most commonly used fire detectors are listed under the title "Thermostats" in the UL Fire Protection Equipment List. They are integral assemblies of heat-responsive elements and noncoded electrical contacts that function automaticially under conditions of rising air temperature. Thermostat types included are:

Fixed-Temperature Fusible Material

This is usually material such as solder that melts to open a circuit or plastic insulation that melts to close a circuit. In one type, a salt becomes conductive when heated above its melting point. Some types have heat collectors to hasten their operation.

Rate-of-Rise

These usually operate on an expansion-of-air principle using a chamber with a small calibrated vent. Normal temperature changes of less than 15° F. per minute will not cause activation, but greater changes will sound an alarm. Because only the rate of change, rising or falling—not the temperature—is important, the units may require special treatment if they are close to heaters or inside or near freezers.

Bimetallic, Fixed-Temperature— Rate Compensated

These operate on the principle that different metals expand at different rates under a temperature change. When the change is sufficient, the difference in length causes contacts to close or open. Most bimetallics appear as a bent spring because the metals are in contact. Others work on a slightly different principle because the materials are not in contact. An important point: The thermal lag or heat absorption rate of each metal causes the detectors to be "rate compensated"—that is, to operate similarly to a rate-of-rise detector. This may be desirable, but if a high-temperature detector is needed and if the rate of temperature change exceeds the thermal lag of the detection material, there may be a false alarm.

Continuous Line Types

Three detectors are in this category. The heat-pneumatic type has tubing circuits that operate on the rate-of-rise principle. The circuits are limited to a maximum of 1,000 feet of continuous-length tubing of

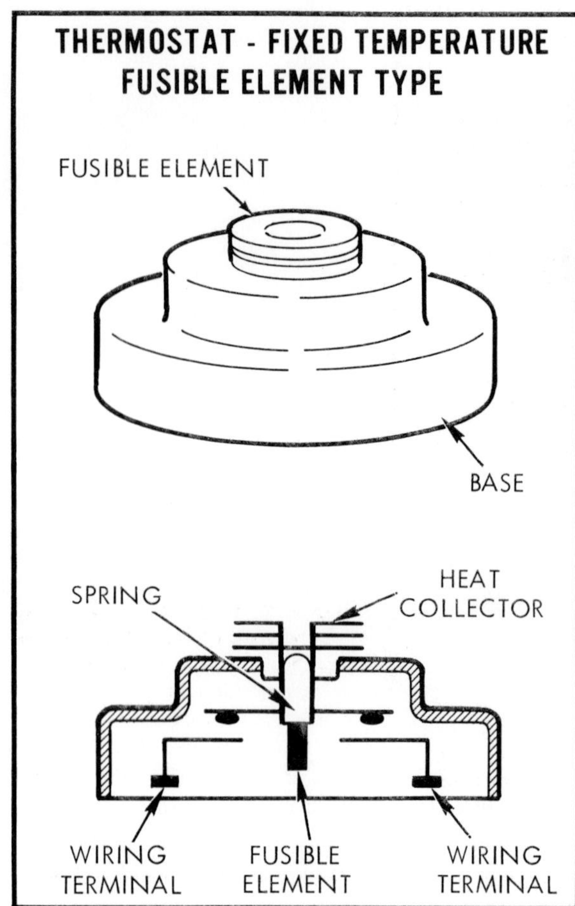

Heat detector usually arranged with normally open (N.O.) contacts.

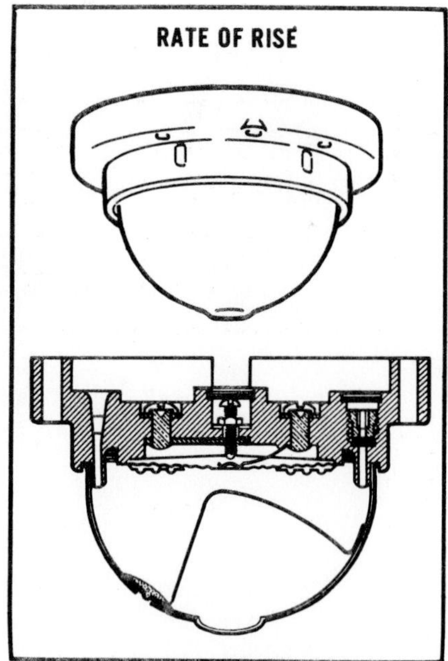

The device shown is a combination rate-of-rise and fixed-temperature type. The fixed-temperature feature is optional. Contacts are normally open. The compensating vent is fixed by design.

about 1/8-inch diameter. The circuits are reliable and relatively easy to maintain and test. They are at a cost disadvantage in small rooms, however, since rules require that at least 5 percent of the circuit must be installed in each room, closet or other area enclosed by walls.

Another type uses two energized spring-steel wires separated by a dielectric thermal plastic. The wires operate on a closed-balanced-loop principle, each wire carrying a small amount of electrical current for supervisory purposes. When heat melts the plastic, the spring-steel wires, twisted around each other in a spiral, come in contact and upset the circuit balance.

A third type of continuous-line thermostat uses a conductor inside a metal tubing jacket separated by a cutectic salt that is dielectric when cool and solid but conductive when heated above its melting point.

Smoke Detectors

These have perhaps the most appeal of all detectors. A conventional smoke detector operates on a light principle; smoke entering a light beam either absorbs light so the receiver end of the circuit registers obscuration, or it scatters light so that a terminal, normally bypassed by the light beam, receives part of the light.

Products-of-Combustion

These are a special smoke detector type and are said to be responsive to both gaseous products of combustion and smoke. These products-of-combustion detectors are advertised as "capable of detecting an incipient fire long before there is visible smoke or fire . . ." The detectors are capable of detecting a fire this small

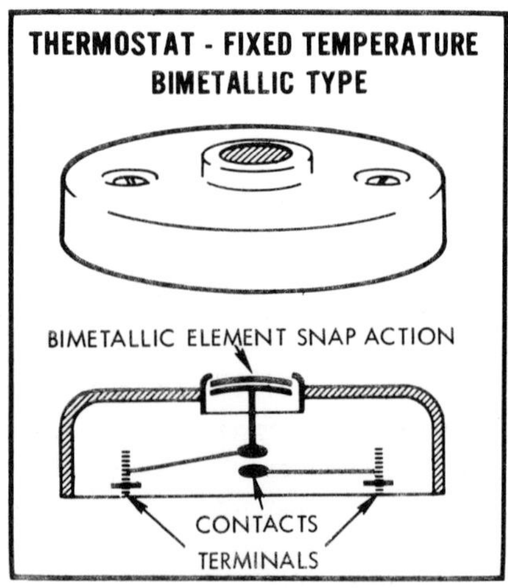

This device may be arranged with normally open or normally closed contacts.

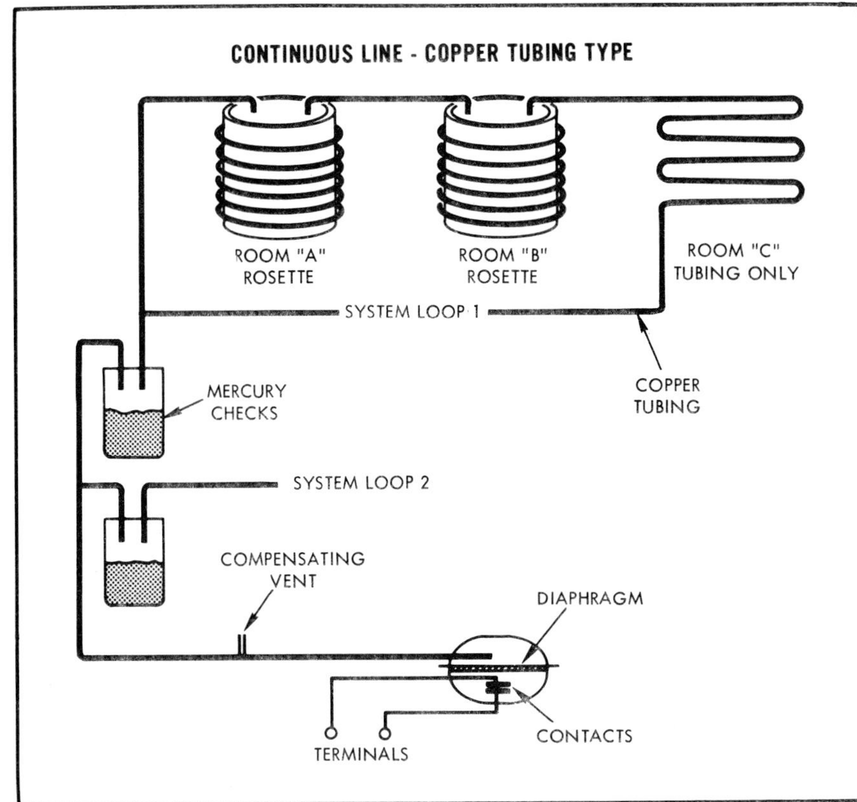

Volume of tubing in a room is critical. Rosettes are used in small rooms. Mercury checks separate system loops. Volume of loops and number of loops are limited. Compensating vent must be set for field conditions. System may have normally open or normally closed contacts.

because they are sensitive to particulate matter between .01 and .1 micron. Visible smoke is usually composed of larger particles, 4 to 5 microns. The catch? The fire must be large enough to deliver enough small particles to trigger the detector. For this reason, it usually requires a larger fire than advertised.

Special Detectors

Other detectors, less frequently used, include:

Flame. These operate on the principle of radiant energy received from a flame. They are primarily line-of-sight devices and must see the flame or its reflection through a special lens. Infrared and ultraviolet units would be similar.

Heat-thermopile. These operate on the principle that two dissimilar metals in intimate contact generate a minute but measurable amount of electrical current when the point of contact is heated. A single circuit is a thermocouple, but those connected in series are thermopiles.

Self-contained single- and multiple-station devices. It is difficult to accept these relatively new devices as systems. The units, including the power supply, are self-contained. The power supply may be electrical, mechanical (spring-wound) or gas-operated. They may be hung on the wall as independent units or connected together in a system. Consult the authority having

jurisdiction before considering these for a protective system.

Sprinkler Supervision

An important part of automatic detection equipment is often not even considered detection. These are extinguishing system attachments. The most important of these are water-flow detectors, which may operate on the principle of a sudden increase or decrease in pressure or by deflection of a vane inside the piping caused by the flow of water. All are designed to detect a flow of approximately 10 gpm.

Supervisory devices. Other attachments monitor functions that indicate a system function is jeopardized.

Water-level devices. These warn when a self-contained water storage supply is low or, in a low-differential dry pipe valve, when the level of water is too high.

Water-temperature switch. This indicates when the storage water supply may freeze.

Water-supply valve-position switches. These signal if someone starts to turn off the water supply.

Power-supply devices. These monitor pump and engine operations and power supply conditions.

Rules for recommending maximum detector spacing and detector locations are shown in the Underwriters

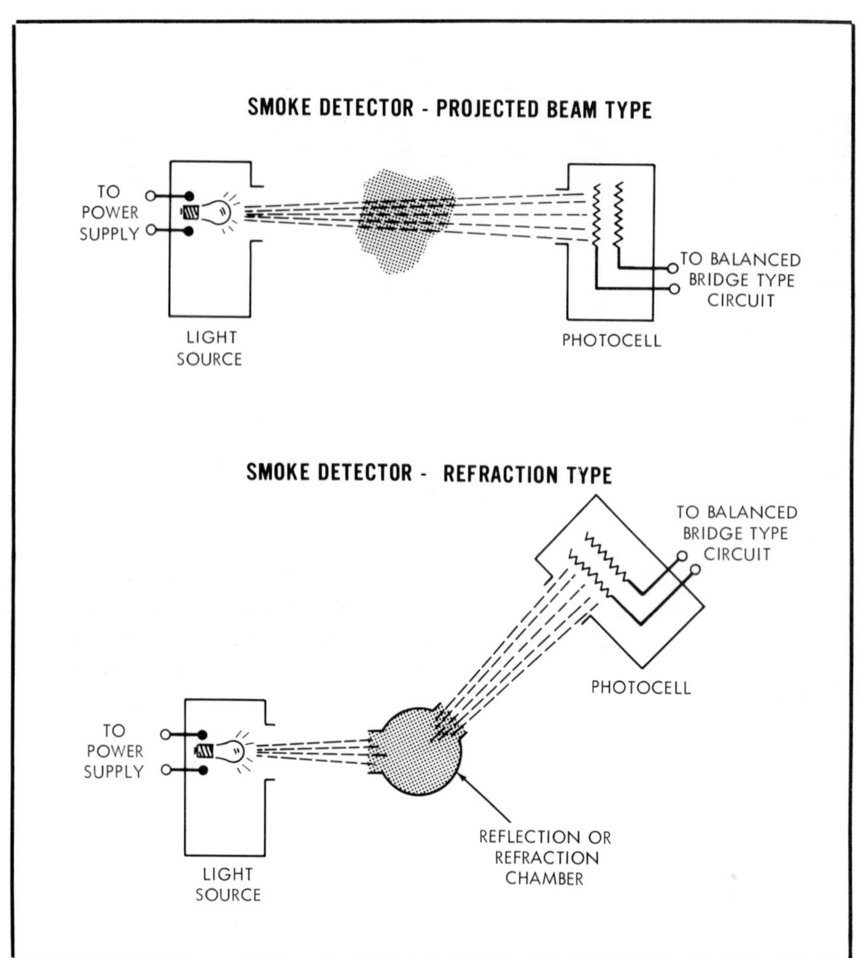

When smoke enters changer, light is reflected or scattered to sensitize photocell. This is a spot-type device.

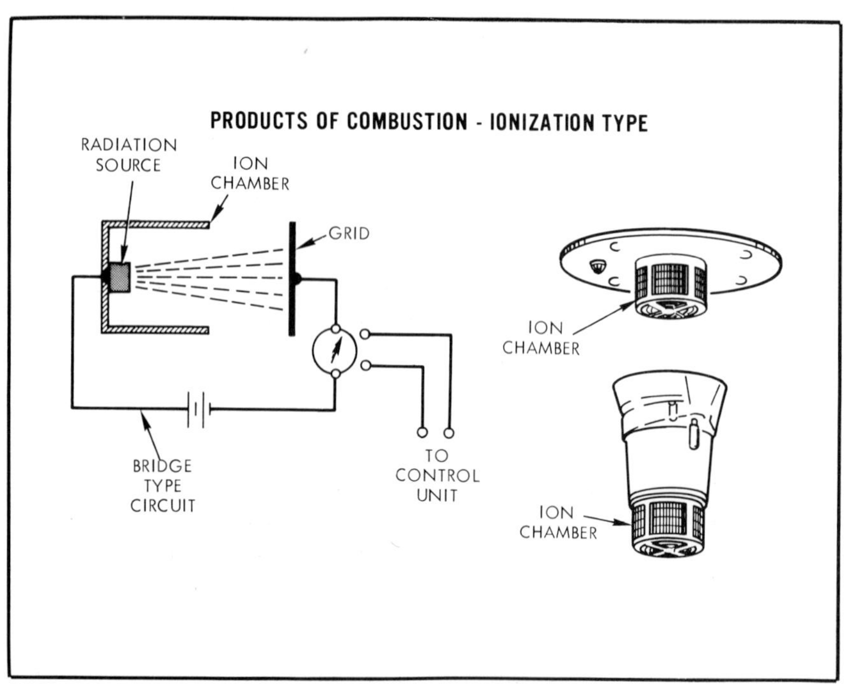

Ion chamber current hrs (normal) 10^{-9} to 10^{-12} amp. Blockage or change in current flow causes operation.

Fire Detector ratings are determined by Underwriters Laboratories, Inc. as "spacings" for heat detectors. Spacing ratings of 15, 20, 25, 30, 40, 50 feet or greater are given based on actual fire tests. The spacing is the distance between detectors installed in a square or box pattern. The conditions of the test tell a great deal about the meaning and application of the spacings.

Figure 1 shows a plan view of the fire test area. The fire is at the midpoint of a square, and the detectors are at the corners. The distance detectors are from the fire is representative of the spacing for which the detector is being tested. This distance is also the maximum distance from the fire that is allowed by proper installation and thus produces the slowest acceptable detector response.

UL FIRE DETECTOR TESTS

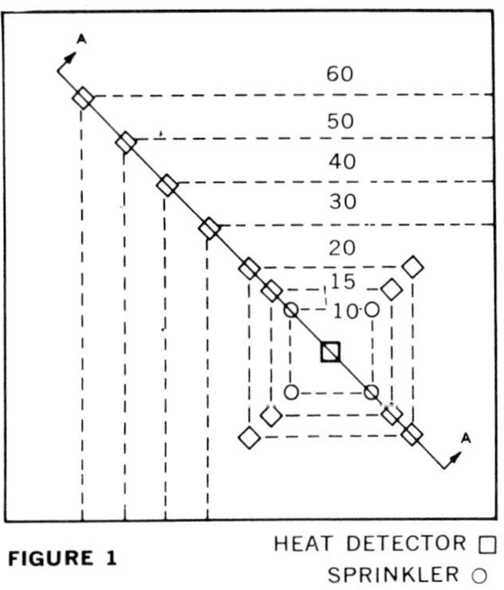

FIGURE 1

Because of space limitations, detectors are at the corner represented by only one diagonal. Automatic sprinklers of fusible link and lever design rated at 160° F. are installed on a 10-foot × 10-foot spacing pattern. They are closed, with

fusible links in place. Figure 2 is an elevation view taken through section A-A. It shows that the sprinklers are located with deflectors seven inches from the ceiling in a room fifteen feet nine inches high. The detectors are mounted directly on the ceiling. The test fire is alcohol, burned in pans three feet above the floor.

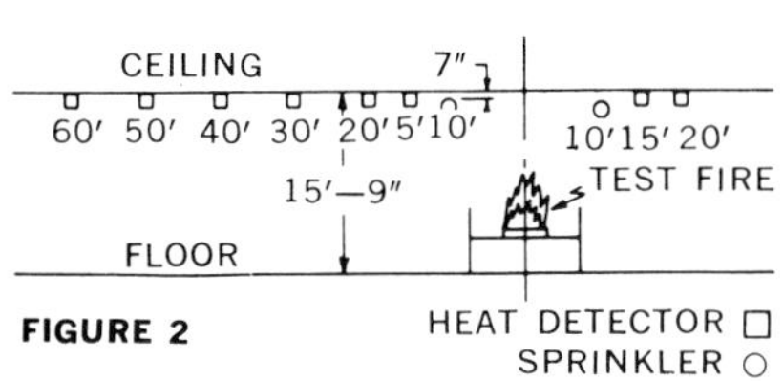

FIGURE 2

Several test fires are conducted with pan fires of various sizes to produce different conditions. In each test, the time of operation of the automatic sprinklers and time of operation of the detectors are measured. To qualify for a spacing rating, the detector must operate before the automatic sprinklers under fire conditions that cause the sprinklers to operate in two minutes, plus or minus ten seconds. In other words, a detector rated for spacing is just slightly faster than an automatic sprinkler installed on a 10-foot × 10-foot spacing when the detector is installed on its maximum spacing.

The philosophy behind this requirement goes back to the need for fire detectors as actuators for supplementary extinguishing systems or as devices that summoned the fire department. The detector must operate before any automatic sprinklers that might be installed in the building if it is to operate at all. After all, when an automatic sprinkler operates, water is discharged, cooling the atmosphere. This will prevent subsequent operation of any detectors. The listing or approval of the detector is based on its ability to sound an alarm or operate a supplementary extinguishing system before sprinklers operate.

Laboratories, Inc.'s Fire Protection Equipment List and the National Fire Protection Assn.'s Signaling Systems Standards. If a fire detection system is to meet only existing codes and standards, these offer adequate guidance. If the engineer wants the best detection for the overall value, however, he must go beyond these publications.

Here are some considerations.

What size of fire must be detected? Detector spacing rules are based on the response sensitivity of a sprinkler head. Thus, a detector installed at maximum spacing may require several minutes to detect a fire involving four pounds of crumpled paper, a relatively large, hot fire. A fire in overstuffed furniture emits much smoke but little heat.

Spacing rules are based on flat smooth ceilings of ordinary height, 15 feet 9 inches (see UL Fire Detector

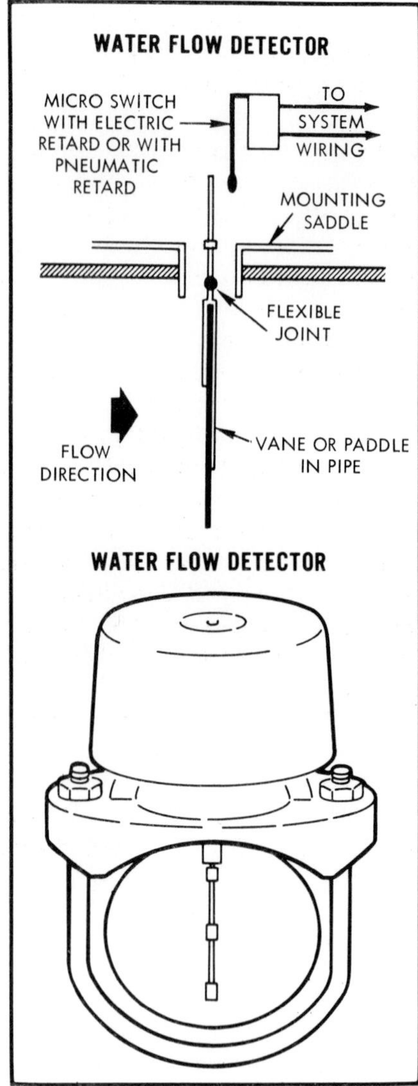

Vane movement actuates switch. Contacts may be normally open or normally closed. Retarding devices are optional.

Tests, above). Detector sensitivity varies as a function of room height. (Experts do not agree, but it appears to vary between the square and the cube of the height.)

Minor obstructions seriously affect the flow of heat across the ceiling, just as they do the flow of liquids across an uneven surface. The detector reaction time might double if it is obstructed by a solid unit that projects only six inches below the ceiling.

It is a good rule of thumb to specify detectors throughout all areas that would require sprinklers if the building is to be sprinklered. Detection should never be omitted from attic spaces or lofts simply because they are inaccessible. The concern should always be: Can a fire start here? Is there fuel to burn?

NFPA standards state that detectors are required below all decks, mezzanines or floor landings that are four feet wide or more. This is based on sprinkler rules and should apply to all horizontal surfaces that could increase the reaction time for the automatic detector.

The standard does not take into consideration the fact that overhead doors in shipping departments often are left open. A fire cannot distinguish between a deck, mezzanine or floor landing more than four feet wide or a ten-foot-wide overhead door or other horizontal obstruction.

Detection devices should be specified to go under steel-grated platforms and floors—even though there are holes in the material. These are often covered with solid materials. Even if the grates are not covered, detectors above the floors would be less sensitive due to height, heat absorption by the material and the restrictions caused by the partial obstruction to heat flow; so baffles should be used over the detectors to collect the heat and aid operation.

Heat collectors are required for sprinklers or for heat-activated devices on preaction or deluge systems but not (according to the Standard) for detectors. To make sure they are included, the engineer should specify them. A suggested design would be a sheet metal collector at least 18 inches $\times$ 18 inches, preferably with a one-inch drop lip, similar to the sprinkler system heat collector. There are, however, acceptable alternatives, such as sheet metal "coolie" hats, obsolete metal light reflectors or even children's metal snow dish-saucers. A device is needed to hold the heat or smoke close to the detector for a quicker operation.

The NFPA standard indicates that detection may be required under outside awning, canopies, or platforms. If material there can burn, detection is desirable unless fire can do no damage to any area of concern.

Detector sensitivity decreases as the distance between detector and ceiling increases. The distance will vary with the type of detector; the threshold for common detectors ranges from about six to twenty inches below the ceiling. Below these limits heat collectors are desirable.

For esthetics, detectors are often put at the intersection of wall and ceiling. This is most often done with continuous-line thermostat detectors. Esthetics aside, manufacturers admit that the area is a "cold pocket," which heat does not readily penetrate by convection currents. The peak of a gabled or sawtooth roof is a similar area where cold pockets exist.

All detectors require a minimum environmental change to activate them. The change could be temperature, rate of change in temperatures, smoke or particulate matter. In any event, a minimum condition at the detector is required, yet detection in large areas is almost invariably spaced regularly, on geometric square intervals. To improve sensitivity, detectors should be arranged in a staggered pattern. Extra de-

tectors should sometimes be installed at points where
heat or smoke may gather.

Detector spacing in corridors is often stretched
to the maximum or even omitted. Spacing in long
corridors should decrease because additional resistance
to air flows is caused by minute pressure differentials,
particularly, in narrow corridors. If there is an ignition source or fuel, detectors should not be omitted
from corridors.

For special hazard areas, where criteria for detection are extremely complicated (special processes,
working atmosphere or others), the best detector may
not be UL-listed. The engineer should consult equipment manufacturers and fire detection specialists for
advice and availability of special detectors.

Can products-of-combustion smoke detection systems be designed? Perhaps, but neither the standards
nor UL tell how to do it. Moreover, the manufacturers
disagree among themselves.

There are devices advertised as being *able* to detect
an incipient fire before there is visible smoke or flame.
The key word is *able.* The detectors are sensitive to
particulate matter normally too small to see (.01 to .1
micron size), but a fire of considerable magnitude may
be required to deliver the necessary particulate in sufficient quantity to trip the detector.

These detectors were originally listed for sixty-foot
spacing (3,600 square feet). However, UL now recommends a thirty-foot spacing (900 square feet), and
even that carries a warning that it is based on test
fires in a laboratory with a 15-foot 9-inch smooth
ceiling, no air movement and no physical obstructions
between the fire source and the detector. The test
fires for the seemingly ideal conditions are detailed in
the UL listing and include: half a pound of shredded
paper, a special pile of wood kindling ignited by 100
cc of alcohol, two-ounce pile of plastic polystyrene
spaghetti packing material, and 200 cc of leaded gasoline. While the detectors are "able" to detect an incipient fire before there is visible smoke or flame, the
test fires cause a good amount of each. The same is
true of conventional smoke detectors: these devices
are not assigned UL-spacings, but systems must be
designed to meet job conditions.

Products-of-combustion detection is particularly
applicable for areas with high-value density or high
susceptibility. They are the best choice where the buildup of flame, heat and smoke will be rapid, because
there is little value in early detection if a second or two
later the fire is roaring. They are of no value if protection—manual or automatic—is not applied promptly.
If the protection is a considerable distance away, the
difference between a fire department response of
fifteen minutes for early warning detectors and six-

teen minutes for conventional detectors may be negligible. Except where hazard to life is the only concern,
early extinguishment is mandatory if there is to be
benefit from early detection.

Where life is the primary concern, in an apartment
building, for example, the problem is different from
that of an industrial plant. In a plant, the concern is
the workers' response to alarms and the subsequent
evacuation. In residential units, the greatest concern
should be with evacuation during sleeping hours.

Conventional detection in a residential situation is
often inadequate because small, smoldering fires do
not generate much heat, flame or smoke. Tests detailed in the January, 1963, NFPA *Quarterly* dealt with
thirteen typical residential fire situations. In the first
test, almost two hours passed before the heat detectors
operated. At that time, the carbon monoxide concentration in the bedroom with the door closed and windows open was 1,670 ppm (a concentration of 1,000
ppm for one hour is dangerous to life). In ten of thirteen
tests, it took conventional detectors more than an
hour to sound alarms.

System cost is secondary to concern for life safety.
As opposed to heat-sensitive detectors, products-of-
combustion detectors appear to be a better solution.
They are not the only answer, however, because effective automatic carbon-monoxide detectors appear to
be nearing a technological breakthrough.

After Design, What?

Designing a system is only part of the picture. A serious
problem today is that equipment is sometimes installed
by persons who quite often do not have adequate
knowledge or understanding of fire alarms. An acceptance test procedure is recommended. This procedure
could include a stability test (a week to a month without unnecessary alarms); sensitivity test (how small a
fire will it detect?); alarm audibility (can it be heard
readily in all areas?); and alarm transmission (can the
receiver understand each function?).

Fire alarm systems are a specialty field, and the
design of systems is best entrusted to those qualified
and regularly engaged in this field.

Where can the engineer go for help? Quite often, it
is best to contact the manufacturer directly, since
he has qualified men who can offer possible solutions
to problems. Consulting firms specializing in fire protection engineering are another valuable source of help.
Nor should the authority having jurisdiction be forgotten. Remember that the engineer in charge of a project doesn't have to have all the answers, but he should
know where to get them.

Electrical Fire Safety

C. F. HEDLUND, P.E.

Hazards associated with electrical systems and equipment are not always clearly covered by the National Electrical Code and are often overlooked by engineers. This chapter describes some of these hazards and common deficiencies in the design, construction and protection of an electric power distribution system and the utilization equipment and ways and means to prevent electrical fire losses by proper application of the rules of the National Electrical Code and other safeguards.

Almost 16 percent of all fires in U.S. buildings during 1973 were of electrical origin, and they caused over 13 percent of the total dollar loss, according to the National Fire Protection Assn. Electrical fires scored second among all known fire causes and ranked first in the amount of loss. In industry, electrical malfunctions cause more fires than any other single agent, reveal the records of Factory Mutual. The percentage of electrical fires remains constant at about 20 percent, but the amount of loss they cause each year varies between 10 and 25 percent of the total fire loss, depending on type of equipment, kind of occupancy and extent of protection.

A five-year study of more than 15,000 fires occurring in Factory-Mutual insured properties shows that 3,370 were of electrical origin and they caused $10.6 million in property damage (not including loss due to business interruption). The study also determined the frequency of fire in the various classes of electrical equipment:

Type of Equipment	Percent of All Electrical Fires
Wiring	29.0
Motors	21.0
Controllers & Switches	12.0
Lamps or Hot Elements	7.0
Vehicle Wiring	6.0
Transformers	2.5
Electrostatic or Hi-freq. Units	2.5
Switchboards	2.6
Generators	1.7
Radio & T.V.	1.3
Circuit Breakers	1.3
Fuses	1.0
Miscellaneous	12.1
Total	100.0

The electrical breakdowns that caused these fires were of different causes, including excessive vibration, overheating because of overloads, failure of cooling equipment, high ambients, plugged ventilation passages, excessive accumulation of foreign materials, and insulation deterioration caused by oil, moisture, abrasive dusts, and corrosive liquids, improper application, poor maintenance and inadequate electrical protection. Improper or inadequate maintenance accounted for 70 percent of these losses.

Adherence to the rules of the National Electrical Code or other properly applied safeguards could have

prevented many of these fires or at least held the
damage to a minimum. Too often, however, the engineer thinks that nothing more is required as long as an
electrical installation complies with the code. The
code actually is the starting point for a safe electrical
installation, as is specifically stated in its introduction
under section 90-1-(b), "This Code contains basic
minimum provisions considered necessary for safety."

Section 90-9(b) continues,

> It is elsewhere provided in this Code that the
> number of wires and circuits confined in a single
> enclosure be varyingly restricted. It is strongly
> recommended that electrical engineers and others
> who are planning installations provide similar re-
> strictions wherever practicable, to the end that the
> effects of breakdowns from short circuits or grounds
> even though resulting fire and similar damage are
> confined to wires, their insulation and enclosures,
> may not involve entire services to premises nor
> interruptions of essential and independent services.

This last recommendation is vitally significant but
has not been given the attention it merits.

Hazard of Grouped Combustible Insulated Conductors

One reason why wiring heads the list of electrical-fire
sources is that conductors with combustible insulation
are often grouped in large numbers in one location,
such as a cable trench, raceway junction box or cable
tray. If fire starts, regardless of cause, the conductors
are usually destroyed, leading to considerable property
loss and extended interruption to normal operations.

A large junction box is commonly installed atop the
full length of a switchboard. All circuits to and from
the switchboard pass through this box. Fires frequently
occur in these locations because of a ground or short
circuit and, before the box covers can be unfastened
and the fire extinguished, the insulation on all conductors within the enclosure is destroyed. The results
of a typical junction box fire are shown in the photos
on page 176.

This typical junction box fire occurred in a Mid-
west plant. The installation, in service about nine
months, consisted of a large motor control center containing ten vertical sections arranged five on a side
and back to back. Each vertical section contained
three to five combination starters in NEMA sizes 1
through 4 to control 37 motors (3-phase, 440-v) of
various sizes from 1 to 75 horsepower. Each con-

troller was in a separate pull-out-type enclosure containing a molded case breaker, contactor, overload
relays and control transformer, with stab-type connections on the back that clipped to vertical bus-bars at
the rear of the controllers. The vertical bus-bars, which
supplied the power to each controller, were in turn
connected to heavier horizontal bus-bars across the
back and near the top of the motor control center.
All branch circuit and control circuit conductors from
each of the motor controllers, consisting of Nos. 12,
10 and 8 wire with type TW, RH & RHW insulation,
passed through the junction box.

The trouble apparently started at the stab-type
contacts at the rear of one of the controllers near the
bottom of one of the vertical rows. Trouble presumably
sprang from a poor contact at one of the stab connections that developed into a three-phase short circuit
or an accumulation of dust and moisture providing a
conducting path between phases at the stab connections.
The feeder breaker, however, failed to open because
the instantaneous trips were set too high to detect the
arcing fault.

The arcing spread progressively to other controllers
near the top of the control center igniting insulation
on the branch circuit wiring and control conductors;
fire spread into the junction box. By the time the side
of the junction box was removed and the fire in the
burning insulation was extinguished all wires were
destroyed. The motor control center was so severely
damaged that it had to be replaced and all branch circuit
wiring had to be renewed. Physical damage neared
$70,000 and plant production halted for about two
weeks.

The damage to the wiring in the junction box could
have been prevented had the conductors been grouped
in small numbers, each group wrapped with asbestos
tape and then painted with a solution of silicate of
soda. The tape normally used for this purpose is 1/16
inch thick and two inches wide and is applied with a
50 percent lap. Special "arc-proofing" tapes are also
commercially available and will afford excellent protection against this hazard.

If wrapping the cables in fire-resistive tapes is not
practical, protect them with an automatic carbon
dioxide system arranged to flood the enclosure containing the wires and cables automatically. This system,
however, is feasible only where the gas can be confined
to maintain the proper concentration to extinguish the
fire.

The insulation of the many approved conductors
used in electrical construction is designated as flame-
retardant or self extinguishing. These designations,
however, are a snare and a delusion as far as their fire

Results of a typical junction box fire. Top photos show damage to grouped combustible control and motor circuit wiring. Bottom photo depicts arc damage at vertical bus-bars at back of panel.

resistance is concerned. The designing engineer should not be misled that such insulations will support combustion. They may be more difficult to ignite than other insulations, but most of them will burn readily when densely grouped and in high temperatures. Many other cases indicate such insulations were completely destroyed by a self supporting fire. All wires in the junction box fire were of the "flame-retardant" type.

Low-Voltage Arcing Faults

Such fires result when engineers' specifications do not comply with code Section 110-10 "Circuit Impedance and Other Characteristics," which reads as follows:

The overcurrent protective devices, the total impedance and other characteristics of the circuit to be protected shall be so selected and coordinated as to permit the circuit protective devices used to clear a fault without the occurrence of extensive damage to the electrical components of the circuit. This fault may be assumed to be between two or more of the circuit conductors; or between any circuit conductor and the grounding conductor or enclosing metal raceway.

In the junction box fire, the feeder breaker was set too high to see the initial fault and severe arcing continued until the break was opened by hand. Although the feeder breaker was rated at 600 amperes, at the time of this loss the series trip coils were set at their 800-ampere rating and the pickups on the instantaneous elements were set at fifteen times or 12,000 amperes. The maximum available fault current at the motor control bus was approximately 9,000 amperes, which would usually be reduced as much as 50 percent by the resistance in the arcing fault. Obviously the pickup settings of the instantaneous trip elements were much too high to protect against this low voltage arcing fault. After the accident, this condition was corrected by reducing the coil rating to 600 amperes and the pickup setting on the instantaneous elements to 7.5 times or 4,500 amperes.

Control Center Design

Another factor that contributed to extensive physical damage involved the design of the motor control center. The area containing the bus-bars extended the full length and height of the central portion of the control center. It was completely open so that ionized gases and electrical arcing were not confined to the point of origin but were permitted to spread throughout the area, thus increasing the damage. If the bus-bars had been enclosed in separate, tight compartments, the physical damage probably would have been greatly reduced.

When specifying switchgear and industrial control equipment, the design of the apparatus must confine ionized gases and electrical arcing to their origin points. Equipment that includes this safety feature will cost more, but the added security against extensive damage and interruption to production will compensate for the expense.

Damage from low-voltage arcing faults in motor control centers is common, and the hazard should be carefully evaluated in all new installations so that the electrical protection provided will afford adequate protection.

Continuous Rigid Cable Supports

Article 318 of the National Electrical Code describes the rules for the proper use of "Continuous Rigid Cable Supports"—also defined as "a unit or an assembly of units or sections, and associated fittings, made of metal or other noncombustible materials forming a continuous rigid structure used to support cables. Continuous rigid cable supports include ladders, troughs, channels and other similar structures."

Cable trays and troughs are used extensively today because they are neat, relatively inexpensive, and space-saving means of carrying many cables from one place to another. In addition, they afford easy access to install additional conductors or replace existing ones. When installed according to code rules, these trays present little hazard. Unfortunately, however, numerous cases indicate that improper use of these trays

results in considerable property loss and an extended interruption to production.

Instead of properly separating the cables as the code requires, they are piled unseparated, many layers deep in the trays, especially control circuits and small motor circuits. If fire starts, the insulation on all conductors is often destroyed and normal operations interrupted.

This hazard is well illustrated by a fire in a board mill where the production line equipment was highly automated. The motor control and protective equipment were concentrated in various motor control centers, several in a single motor control center room. Control cables and power cables to the various motors were installed both inside and outside the room in open ventilated cable trays stacked in some places as many as five tiers high. (See photo on page 178.)

The production line was powered by eighty motors ranging from one to ten hp, all arranged for simultaneous operation. A short circuit developed in the conductors to a two-hp motor. The fire started in the control center and spread to the control wires near the ceiling of the control center room. The initial fire increased the temperature at the ceiling, where the trays were— hastening the spread of the fire. Approximately 1,000 cables of three to nineteen conductors per cable were destroyed and the metal deck on the bar joist roof was so seriously damaged it had to be replaced. The motor control equipment in the room also suffered extensive damage from fire, smoke, and water. The damage totaled hundreds of thousands of dollars and the plant shut-down lasted six weeks.

In addition to applying the code rules for "Continuous Rigid Cable Supports," to adequately protect a cable tray installation all wires or cables in the tray should have a fire resistant insulation or jacket. Do not be misled by the terms "flame retardant" or "self extinguishing" as applied to insulating materials. These may ignite less easily than other insulations, but most of them will burn readily when densely grouped and under elevated temperatures.

In a large installation involving numerous long runs of cable trays containing many conductors, the only practical way to protect them against fire may be to install a line of sprinklers over them. If the trays are in a confined area, an automatic room-flooding carbon dioxide system will afford good protection but an automatic sprinkler system should be used as a back-up, to protect against reignition of combustibles after the gas has been discharged. This back-up protection is necessary because carbon dioxide gas has little cooling effect and smoldering particles or metal parts, heated by electrical arcing or excessive current, are sometimes hot enough to reignite after the gas concentration has dissipated.

Damaged control cables and power cables were installed inside and outside motor control center room in ventilated cable trays.

It may be desirable to install automatic smoke detectors over the cable trays and connect them to an alarm system in a constantly attended location. The alarm will sound before a fire has a chance to develop. The fire, therefore, can be extinguished with hand equipment, minimizing damage.

Fire Pump Circuits

When property protection demands an electric motor-driven fire pump, the electrical supply and control equipment must be properly installed. The electrical protection required for the fire pump circuit and motor differs from the usual motor installation, though the reason for this is not always understood.

To insure continued operation of the fire pump especially during a fire, individual short circuit protection of the fire pump feeder is not required, and the running overcurrent protection ususally provided for other motors of the same size is omitted. The National Electrical Code permits such installation in the section headed "Motor Running Overcurrent (Overload) Protection" under paragraph 430–31 (b): "These provisions shall not be interpreted as requiring overcurrent protection where it might introduce additional or increased hazards as in the case of fire pumps. (See NFPA Standard for Centrifugal Fire Pumps No. 20.)"

Ordinarily the code requires each continuous-duty motor rated more than one hp to be protected against running overcurrent by a separate device that is responsive to motor current. This device must trip at not more than 125 percent of the full load current rating of the motor if the motor has a temperature rise of not more than 40° C., but it must trip at not

over 115 percent for all other types of motors. This running protection, however, is not required or desired for a fire pump motor, because the pump and its driving motor are supplied as a unit and the motor is so selected that under maximum pump discharge the motor cannot be overloaded. Foreign material, such as sticks, stones, and gravel jamming the impeller, or a seized bearing may stall the motor. Therefore, the circuit breaker in an approved controller must trip in at least twenty seconds to prevent the motor from burning up because of stalled rotor currents and to protect against a short circuit occurring anywhere on the load side of the circuit breaker. Some commercial types of overload relays have failed violently under short circuit conditions, causing severe damage to the controller. To eliminate this potential hazard, overload relays are omitted.

Figure 22.1 shows the major components and typical arrangement of an acceptable fire pump circuit and controller. The feed is taken directly off the secondary side of a transformer but ahead of the main secondary breaker so that power to the plant may be shut off without affecting the supply to the fire pump. Overcurrent protection for the feeder is usually omitted as one possible source of trouble that may temporarily interfere with the operation of the fire pump because of premature operation, overheating caused by poor contacts, mechanical damage, lightning or other unforeseen hazards.

If, however, local authorities insist on separate overcurrent protection for the fire pump feeder, a circuit breaker should be used instead of fuses, because fuses cannot always be quickly replaced when one or more of them operates. If a fuse operates before or during a fire, the pump is out of service until the fuse is replaced. Replacement requires not only the presence of a qualified person but also spare fuses of the correct size—both are often missing. In the meantime the property may be destroyed for want of a fuse or someone who knows how to locate and replace the defective one. If a circuit breaker trips prematurely, it requires no special training to close it and restore power.

The circuit breaker selected to protect the feeder should be set to trip for short circuit conditions only: about ten to twelve times the full load current of the motor. In addition, its interrupting capacity must be adequate to handle the short-circuit current available from the power supply. If the feeder is connected directly off the secondary side of a transformer, the rating and impedance of the transformer are the principal factors governing amount of short-circuit current available.

If, however, power is supplied directly off a public utility system, only the utility company can furnish

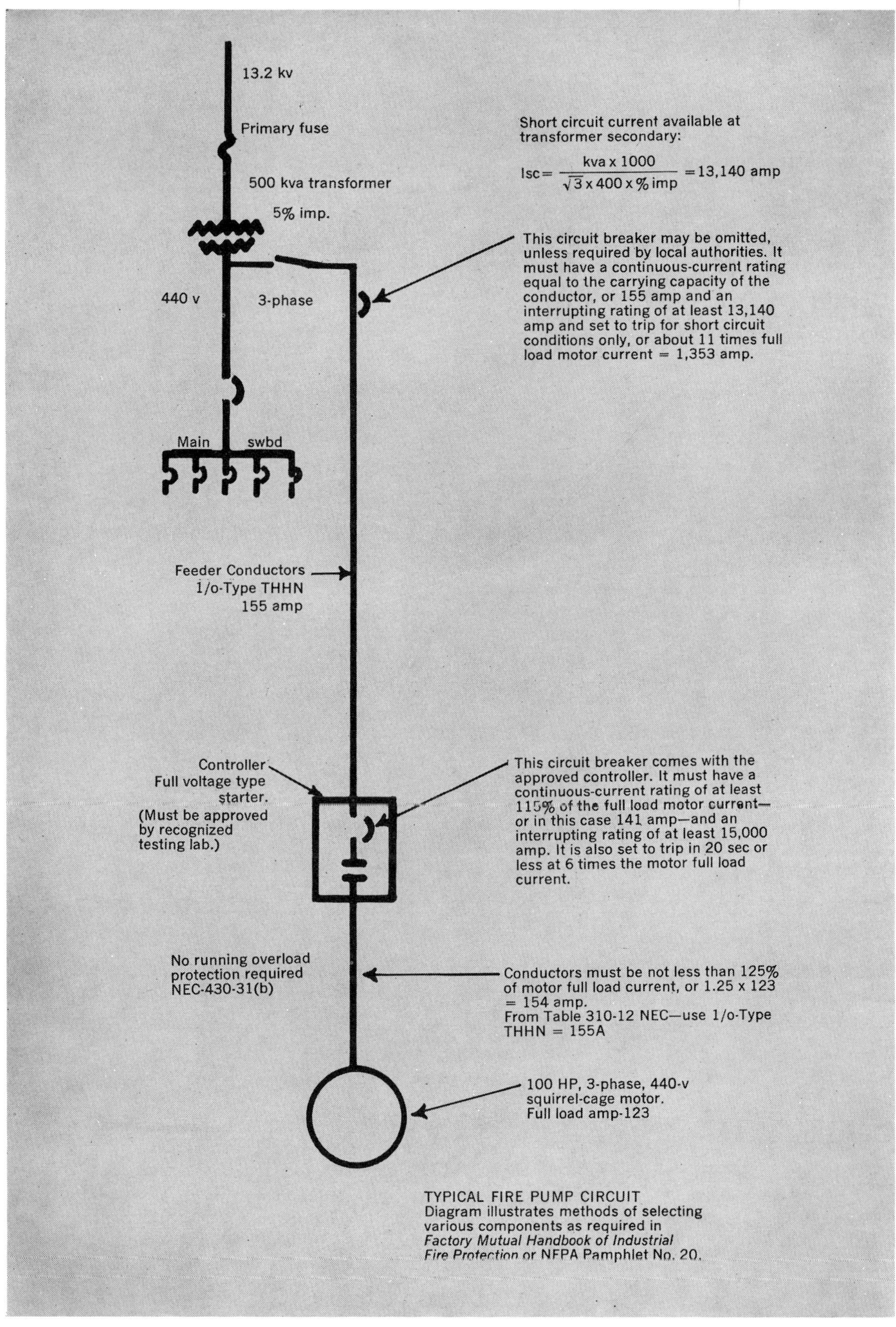

TYPICAL FIRE PUMP CIRCUIT
Diagram illustrates methods of selecting
various components as required in
*Factory Mutual Handbook of Industrial
Fire Protection* or NFPA Pamphlet No. 20.

Figure 22.1.

needed information on the short-circuit current available, in order to select a circuit breaker that has adequate interrupting capacity for the conditions. Regardless of conditions, the circuit breaker in the approved controllers must also have a minimum interrupting capacity of 15,000 amperes. The feeder to the fire pump cannot be located where it is exposed to unprotected combustible structures or occupancies. If the feeder is exposed to possible lightning damage suitable lightning arresters should be provided on the lines where they enter the building.

High-Rise Hazards

Bus-ducts extending from the service equipment in the basement up a vertical shaft to the top story in a high-rise building occasionally experience fire.

In one case, fire completely destroyed an eighteen-story ventilated low-impedance bus-duct rated at 4,000 amperes, 600 volt, three-phase, three-wire. This bus-duct was constructed of twelve aluminum bus-bars with four bars per phase. Each phase was wrapped with varnish cambric tape to a thickness of 20 mils and was supported every foot in the metal enclosure on ebony asbestos insulators. Hair felt 1/4 inch thick was also installed between each ebony asbestos bus support and the sides of the bus-duct and two 1/16 inch thick fiber strips were used to separate the bus-bars of opposite phase.

The fire started near the bottom of the bus-duct and spread up its full length. Before the fire was extinguished, the duct was so severely damaged that complete replacement was necessary.

A similar fire occurred in the same kind of bus-duct in a ten-story building. In this case, a phase to ground fault occurred during construction and caused the fire. Here again the bus-duct was so extensively damaged that complete replacement was required.

These fires emphasize the need to eliminate, in so far as possible, the use of combustible materials in bus-duct construction. Where significant quantities are used, as in the foregoing cases, substantial fire-stops should be provided about every ten feet in the bus-duct.

Municipal inspection authorities report that many new high-rise apartments are supplied from transformers that have secondaries connected directly into power panels on each floor with no means of de-energizing except at the primary disconnects that are frequently inaccessible to the tenants or the property owners.

No information indicates that this deficiency has contriubted to any serious losses, yet the potential hazard exists. Many persons familiar with the code disagree as to whether it requires a suitable disconnecting means for the foregoing conditions. Nevertheless, it is good practice, when laying out the protection for high-rise buildings, to be sure that a load-break disconnecting switch or a circuit breaker is accessible to the occupants so that any power or lighting panel may be quickly de-energized.

Fluorescent Lighting Equipment

Electrical breakdowns in the ballasts of fluorescent lighting fixtures have also caused numerous fires. Overheating of the ballast occurs for several reasons and failure usually follows. When inadequately protected, molten compound may drip out of the fixture, sparks may flare, combustible gases from the overheated compound may explode, or a combustible ceiling may be ignited.

Section P of Article 410 of the National Electrical Code outlines the special provisions for fluorescent lighting systems.

The ballasts furnished with the fixture, therefore, must now have a built-in thermostat that will open the circuit if the temperature of the ballast reaches 120° C. When the ballast cools down, the thermostat recloses, but it does not remove the cause of the overheating.

For complete protection, in addition to the inherent overheating protection of the ballast, the ballast in each fluorescent fixture should be individually protected with a small fuse, depending upon the wattage and voltage of the unit.

Intrinsically Safe Electrical Equipment

In areas where volatile flammable liquids or flammable gases are handled, processed or used or where combustible dusts are in suspension continuously, intermittently or periodically, electrical equipment that produces an arc or spark during its normal operation must be housed in explosion-proof enclosures approved for the specific gas or dust.

The electrical installation must also conform to the special requirements outlined in Articles 500, 501 and 502 of the National Electrical Code. Complying with these requirements always costs much more than an electrical installation in a location of ordinary hazard. The use of intrinsically safe electrical equipment with ordinary electric wiring in a hazardous location has evoked increasing interest among system designers.

However, everyone does not clearly understand the concept of intrinsically safe electrical equipment. In addition, the supply of intrinsically safe equipment or devices that have been approved or listed by a recognized testing laboratory is limited. The occasions when an intrinsically safe electrical installation may be substituted for the type normally required for hazardous locations are, therefore, very limited.

Intrinsically safe equipment and wiring is defined in Article 500-1 of the code as "incapable of releasing sufficient electrical energy under normal or abnormal conditions to cause ignition of a specific hazardous atmospheric mixture. Abnormal conditions will include accidental damage to any part of the equipment or wiring, insulation or other failure of electrical components, application of over-voltage, adjustment and maintenance operations, and other similar conditions."

Approval standards have recently been published by the Factory Mutual Engineering Corp. specifying the requirements that must be met for a device to be considered intrinsically safe. The Instrument Society of America has also published a tentative recommended practice for intrinsically safe electrical instruments. Both documents introduce the concept of "non-incendivity," which applies to equipment suitable for Class 1 Division 2 *hazardous locations, as refined in National Electrical Code Article 500.* The term "non-incendive" is defined as equipment that "in its normal operating condition would not ignite a specific hazardous atmosphere in its most easily ignited concentration." (Emphasis added.)

Intrinsically safe and non-incendive equipment differ in that the latter is not required to be safe under *abnormal* conditions and is only suitable for Division 2 locations. *Normal* conditions, however, would include maximum supply voltage and extreme environmental conditions within the stated ratings of the equipment.

Fundamentally, an intrinsically safe device must meet the following requirements:

- The maximum available energies in the electrical circuit under abnormal conditions must be less than the minimum ignition energy of the specified hazardous material that is involved. Abnormal conditions include any two mechanical or electrical faults or failures that are independent of each other, and not the direct result of one another, that have a reasonable probability of occurring in combination.
- Intrinsically safe parts of the electrical device must be safe under fault conditions in the interconnected portions of the device. This safety requirement applies to the field wiring that connects the intrinsically safe equipment to those parts of the equipment in non-hazardous areas. It also applies to those components located in the non-hazardous areas and commonly used to limit the release of energy to an amount below that required for ignition.
- The means of limiting the maximum available energy release must be effective, practical and reliable. This end is usually accomplished by means of snubbing devices such as capacitors, resistors and semi-conductor rectifiers that will hold the energy within safe limits. The slow operation of fuses and circuit breakers makes them unacceptable for limiting the energy. Another method of limiting the energy release is by encapsulation of some of the parts to protect them from accidental short circuits and to insure that any energy released from an encapsulated part of the circuit will be through the desired circuit impedance.
- Calculations to determine the maximum energy available must assume maximum power supply voltage, ambient temperatures of $0°$ to $120°$ F, atmospheric pressure, and all likely pairs of fault conditions.
- Equipment must be properly designed for the intended application and only high quality components used and applied at no more than 50 percent of their electrical rating.
- If calculations indicate that the maximum energy available is a significant fraction of the minimum ignition energy required for a particular hazardous material, the device must be subjected to ignition testing to show that a safety margin actually exists.

Though presently only a small amount of equipment has been approved as intrinsically safe by any recognized testing laboratory, the electrical designer should consider employing some of these devices in those areas that otherwise would require the use of explosion-proof equipment. Intrinsically safe equipment now available has been approved to sense, control and record pressurized fluid processes, for detecting gas leaks, actuating pneumatic controls and remote electronic controllers for contactors.

References

National Electrical Code—National Fire Protection Assn., 470 Atlantic Ave., Boston (NFPA 70-1970)

Factory Mutual Handbook of Industrial Fire Protection—Chapter 34-14, Fire Hazard Study, Grouped Electrical Cables, NFPA publication

Arc-proofing Tapes—3 M Co., St. Paul, Minn., Johns Manville Co.

Grouped Combustible Cables—C. F. Hedlund, Fire Journal, NFPA, March, 1966.

The Problem of Arcing Faults in Low Voltage Power

Distribution Systems–Francis J. Shields, *IEEE Transactions, Industry and General Applications, Jan./Feb.,* 1967, Vol. IGA3, No. 1

A New Ground Fault Protective System for Electrical Distribution Circuits–Richard R. Conrad, Derio Dalasta. *IEEE Transactions, Industry and General Applications,* May/June, 1967, Vol. IGA3, No. 3

Fire Pump Controllers, For Electric Motor Driven Pumps, Factory Mutual Approved Guide, Factory Mutual Engineering Corp., 1151 Boston Providence Turnpike, Norwood, Mass.

Pumping Equipment For Fire Service–Underwriters Laboratories, Inc., Fire Protection Equipment List

Intrinsically Safe Electrical Equipment–Underwriters Laboratories, Inc.–list equipment relating to hazardous locations, Industrial Control Equipment 184N13.1.

23

Lightning Protection Systems

ROBERT D. HARGER

A typical lightning protection system consists of air terminal and ground electrodes, plus low-resistance conductors to connect the two. Sound simple? Deceptively so. Like most building systems, lightning protection systems call for sound engineering and knowledge of certain facts. Underlying this should be a healthy respect for the awesome destructive power of lightning.

The average lightning discharge is a 200-million-volt, 30,000-ampere shot that can occur in a 30-millionth of a second and reduce a building to rubble. Obviously, a lightning protection system is one of the most vital systems in a building. It is imperative that such a system be designed, specified and installed properly.

Hot and Cold Bolts

There are two kinds of lightning discharge: hot and cold. The difference is related to the power dissipated in the lightning discharge. When the level of power exceeds that needed to ignite combustibles, the bolt is called a hot bolt. This is usually characterized by a higher than average current, voltage or duration. Cold bolts have relatively low current or duration. Both usually produce explosive effects and both may be single or multiple strokes. The explosive effect should not be confused with the shattering of tree trunks. While the cold bolt will physically shatter a tree, often the hot bolt causes shattering by raising the temperature of moisture entrapped in the cabmium layer to the point where a hydrostatic explosion takes place.

It is not difficult to understand how fires are caused by short circuiting of ordinary house wiring systems, for many of us have witnessed these. Lightning is quite another matter, however. While lightning is an electric current, its energy level is difficult for most to comprehend. The level of current in a lightning bolt is 200 to 2,000 times the level of most common circuits. The voltage pressure of a single lightning discharge is 10,000 to 100,000 times that of common circuit voltage. It has been said that a single lightning bolt can instantly deliver enough power to lift the *S.S. United States* six feet into the air. Regardless, just as normal electric current is carried safely over a system of low-resistant conductors, the tremendous force of lightning can also be carried safely over a system of air terminals, low-resistance conductors and ground electrodes.

Purpose of Lightning Protection

The lightning protection system is designed and installed to carry lightning discharges safely to ground and/or cloud. Although the leader strokes travel from the cloud to make contact with the lightning rod air terminal, tree or building projection, the major current flow is normally from the earth (positive) to the cloud (negative).

The interconnection of metallic bodies in or about a structure eliminates the possibility of flashover due to difference of potential of such bodies. Similarly, grounding mediums, such as those for electrical and telephone systems, water pipes and gas pipes, are interconnected to provide common grounding and reduce flashover.

Lightning protection systems can be installed exposed on the exterior of a building, semiconcealed or fully concealed within the structure. The basic requirements for all systems are the same. In semiconcealed systems, roof conductors are exposed and the down-lead cables and ground connections are concealed within the structure. In concealed systems conductors may be coursed under roofing materials, under roofs, behind column facings or exterior wall facings, or in between wall studs.

Groundings may be made inside buildings under basement floor slabs, but the preference is to carry the conductors to the outside of foundation walls. Ground connections must always be made in soil that will normally be moist. If it is necessary or advisable to make the ground connection inside the building foundations, care should be taken to assure that the ground electrodes are driven deep enough below the slab to assure their being placed in permanent moisture or a counterpoise loop must be installed below the slab.

Chimney points and conductors can be built into the masonry of the chimney. It is important to place such conductors behind the facing materials of the chimney rather than between the flues and the structural masonry. This protects the conductors from the corrosive gases or air mixtures that may penetrate flue tile joints and attack the conductors.

Often, ferrous metal conduit will be specified for use as protection of lightning conductors in concealed systems. This should be avoided because of the difference of magnetic flux about the ferrous metal pipe and that along the nonferrous conductor, when conducting a lightning discharge. This difference of magnetic fields has been known to produce a "choking effect" which can cause the severance of the conductor. To minimize this effect, installation standards call for the positive bonding of conductors to ferrous metal pipes where they enter and leave such pipes. Ridged plastic pipe is recommended where pipe chases are required.

The height of buildings will dictate the sizing of conductors and other equipment required. Basically, buildings over seventy-five feet in height will require Class II, heavy-duty conductors and equipment, while buildings less than seventy-five feet in height will require Class I, or ordinary type conductor and equipment. Special conductors and components are required for heavy-duty smoke stacks.

Lightning protection systems will always have at least two earth (ground) electrodes and ground conductors spaced as far apart as practicable. Air terminals are usually installed on a closed-loop roof conductor and spaced at or within two feet of roof edges, corners of chimneys and other roof projections, and at intervals of twenty feet along main ridges or around the perimeter of flat roofs. Secondary bondings are made to all metallic bodies of appreciable size, such as gutters, downspouts, metal ventilators, soil pipes, etc., if located within six feet of the lightning protection system.

The exact number of ground electrodes as well as the number of air terminals is dependent on the size and design of the structure to be protected. Reference should be made to one of the lightning protection codes for the proper number of spacings.

Systems for Reinforced Concrete Buildings

Protection for reinforced concrete buildings is most easily installed during the construction of the building. Only at this time can proper interconnection of the reinforcing steel and the lightning protection down-lead cables be completed. Adequate connections are made to the lightning conductor system at the highest and lowest points of the reinforcing steel. The reinforcing steel must not be used as the lightning conductor unless such steel is made electrically continuous by welding

---

The Damage That Lightning Can Do

Lightning is the most destructive force in nature. In the United States, lightning causes more loss of life and property than floods, tornadoes, hail and wind storms. Even with more extensive use of lightning protection systems in recent years, losses caused by lightning have increased because of the greater number of buildings and the growth in population and consequent increase in population density.

Annually, lightning kills 400 to 600 persons and injures 1,000 to 1,600 in the United States. A staggering 49 percent of these casualties occur in unprotected buildings.

Lightning Damage in the United States in 1967

Type of Structure	No. Damaged or Destroyed	Dollar Loss
Industrial-Commercial	1,615	$ 99,425,000
Institutional buildings	963	43,266,000
Miscellaneous buildings		20,000,000
Residences (non-farm)	9,004	56,907,000
Farm properties	1,300	52,471,000
Power lines, trees, etc. (est.)		50,000,000
		$322,069,000

and is of adequate size. Normal wire ties will not necessarily render the steel electrically continuous. Special attention must also be given to metal piping systems within reinforced concrete buildings, particularly if they are of extended length.

Marina City Towers in Chicago is a good example of lightning protection in a reinforced concrete structure. The building is sixty-five stories (588 feet) high. The design required special attention to features not found in the ordinary high-rise building. A UL "Master Labeled" system was installed. Six equally spaced nickel-plated bronze air terminals were installed on each of the thirty-five foot diameter penthouse roof parapets. All are connected to a copper conductor within the concrete parapet, with additional bonding conductors to metal housings of mechanical equipment on the penthouse roofs.

Additional air terminals are located around the periphery of the main roof at the sixty-first floor. These are looped together with a copper roof conductor. The conductor in the penthouse parapet is connected to this conductor by means of four copper down-lead cables coursed in the concrete penthouse walls.

Eight of the sixteen periphery building columns of each tower contain heavy wall iron pipe downspouts to drain the main roof and the balcony floors. The penthouse and sixty-first floor conductors are connected to these downspouts, which serve as down conductors to the underside of the twentieth floor.

From the twentieth floor down, the column reinforcing steel is butt-welded end to end to form a continuous metal path down to the underground plates on top of the foundation caissons. At the twentieth floor the downspouts are bonded to the reinforcing steel. The result is that the downspouts and reinforcing steel provide eight metal down-lead conductors from the roof level to the metal plates of the foundation caissons.

The caisson top metal plates at each tower are bonded to an underground copper counterpoise cable. The counterpoise conductor for each tower is connected to eight driven ground electrodes, ten feet long, and the two counterpoise conductors are bonded together with an interconnecting conductor. In addition, the ground counterpoise system is connected underground to the main cold water pipe ground bus.

The many exposed metal balcony railings and balcony-divider screens on Marina City presented an additional lightning problem. There were two dangers: lightning discharge to the balcony railings at the higher elevations and the more serious consideration of secondary or induced currents on the metal railings. For example, lightning discharge to a nearby structure could induce currents on the railings and/or divider screens. Such currents would be seeking a neutralizing path to earth. Paths for these currents were provided by bonding each balcony railing and divider screen to the lightning protection system through the balcony floor drain pipes. All connections are buried in the concrete balcony floors.

Throughout the towers the electrical grounding system is bonded to the lightning protection down-lead system in order to minimize lightning induced electrical system surges. The tower elevator rails, plumbing riser pipes, and electrical riser conduits are bonded to this system at several points throughout the height of the towers to achieve a minimum of electrical impedance in the metallic conducting paths.

This completely integrated lightning protection system provides a well-grounded metal cage around the tower structures to give protection against side strokes as well as normal "top" lightning discharges.

Structural Steel Systems

In structural steel buildings, the steel columns of the building can be used as down-leads. Generally, every other column is grounded with the appropriate ground

electrode. In high-rise buildings, however, it is good
engineering practice to ground all columns and to
connect all ground electrodes with a buried counterpoise cable.

Roof air terminals should be connected to one
another by a roof conductor that is in turn bonded to
the structural steel columns at 100-foot intervals or less.

On low-rise steel buildings, secondary bondings to
bodies of induction are not always required, because
many of these bodies are inherently bonded to the
building steel.

For high-rise steel buildings, the engineer must be
concerned with high-resistance joints in metallic piping
systems and the poorer electrical conductivity of the
metals. The result is possible high-impedance paths.
To overcome this problem, it is recommended that a
ground loop conductor be installed at certain levels
throughout the height of the structure. Number and
location of these installations are dictated in part by
the design of the structure.

For example, the John Hancock Center building in
Chicago has four ground loops located at each of the
four mechanical equipment floors. These are connected
to each of the outside columns and to all main electrical,
plumbing, air conditioning and fire protection risers.
They are installed in addition to the ground counterpoise system and the encircling roof conductor.

Pre-Cast Concrete Units

Many buildings are designed today using pre-cast concrete units, i.e., copings, column covers, curtain walls,
window mullions, etc. They contain varying amounts
of reinforcing steel. Every effort should be made to
coordinate the fabrication and installation of the concrete units so that the reinforcing steel will be electrically
connected to the lightning protection system.

It is a relatively simple matter to weld the tie-plates
or anchor plates to the reinforcing steel before the
units are poured. On installation, a simple connection
to anchor or tie-plate will insure that the steel in the
unit is placed at the same electrical potential as the
lightning protection system. Lightning damage caused
by the failure to properly interconnect such units may
be unusually expensive where repairs must be made
ten, twenty or forty stories above the ground level.

Types of Materials

Copper and aluminum lightning protection materials
are regularly inspected and labeled by Underwriters
Laboratories. Under special conditions other materials,

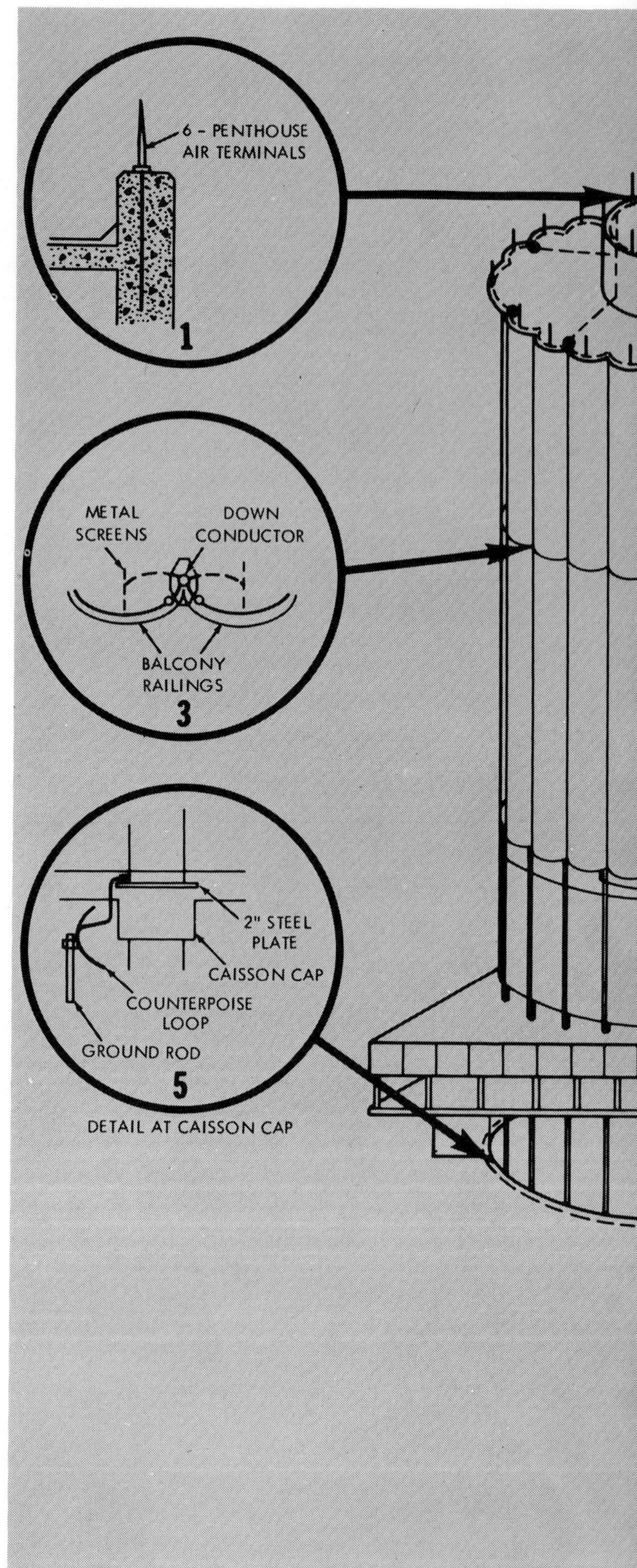

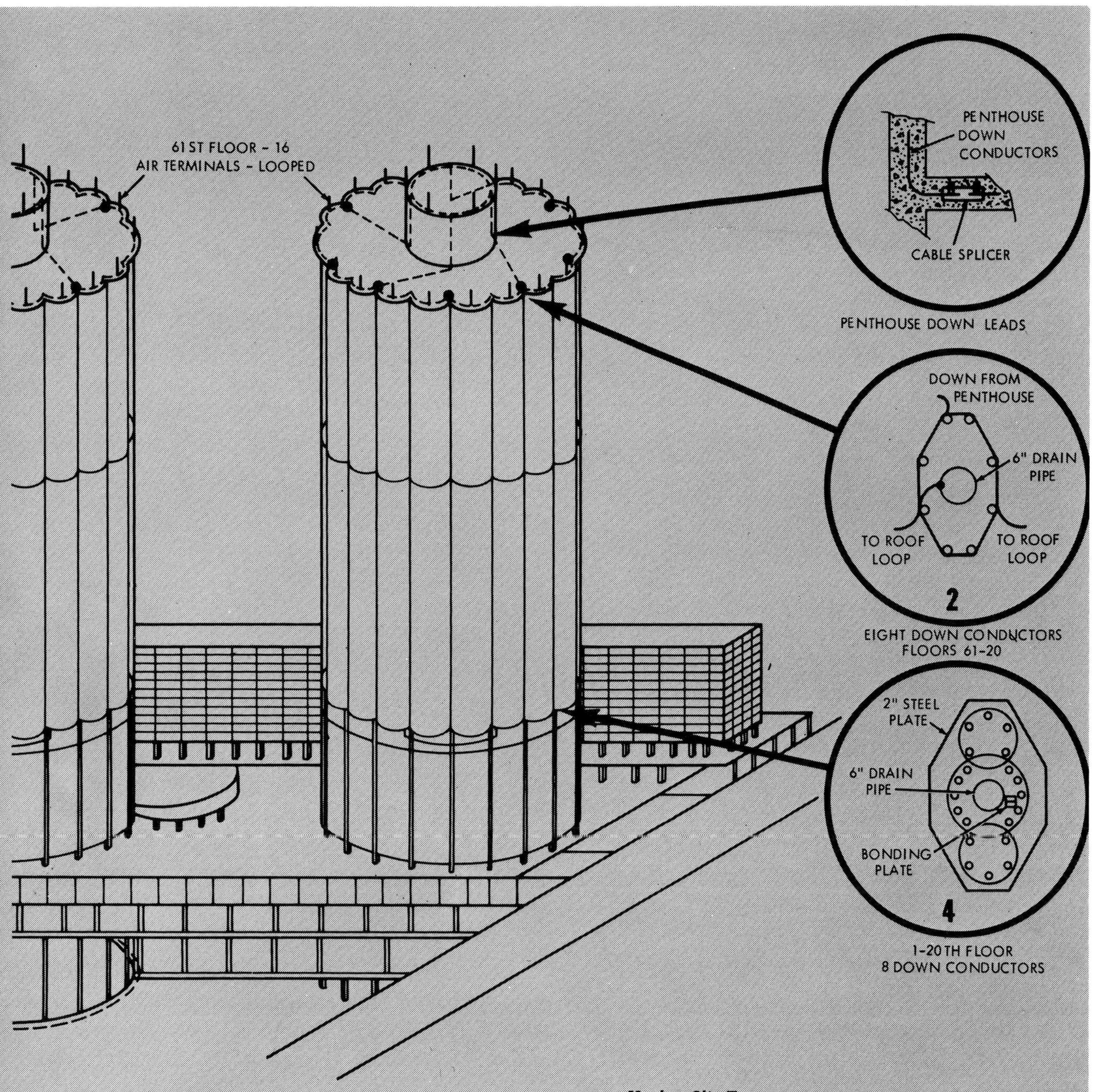

Marina City Towers

SCHEMATIC DIAGRAM OF LIGHTNING PROTECTION SYSTEM FOR A REINFORCED CONCRETE BUILDING
1. Air terminals and roof conductor. 2. Roof conductor bonded to drainpipe and reinforcing bars used as down conductors. 3. Railings and screens bonded to ground conductors to prevent damage from induced charges. 4. Intermediate connections at 20th floor level. 5. Counterpoise conductors with ground electrodes located below ground water level to assure adequate and uniform dissipation of the charge.

which are determined by special investigation to be equivalent to "approved" materials, may be accepted. For example, for smoke stacks having extremely high temperatures at the top, which would melt the lead coverings on regular copper stack air terminals, stainless steel or monel metal air terminals are specified. Caution should be exercised to avoid the use of dissimilar metals which may form an electrolytic couple that might result in the displacement of one or the other of the metals.

How to Specify

Perhaps the mistake made most often is the specifying of "a lightning rod." It is the rare system that would consist of a single rod and still provide the needed

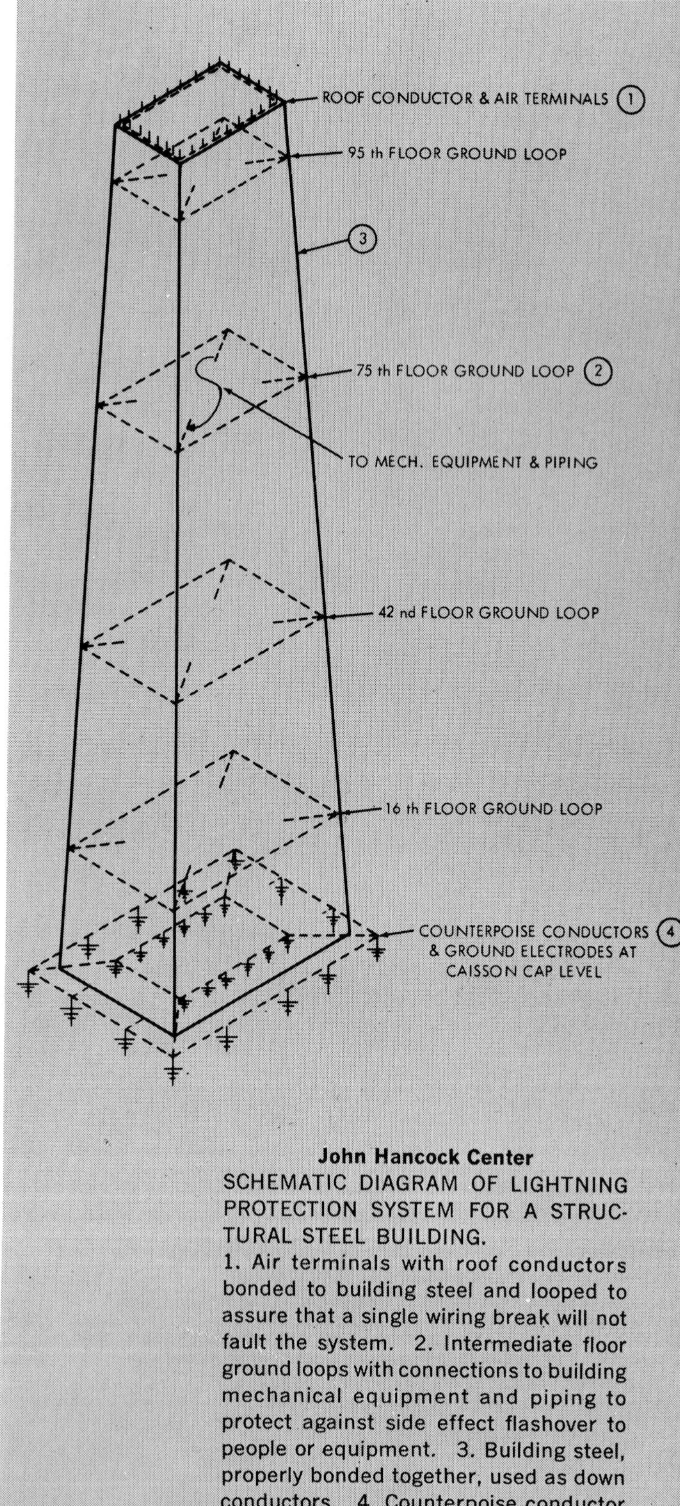

John Hancock Center
SCHEMATIC DIAGRAM OF LIGHTNING PROTECTION SYSTEM FOR A STRUCTURAL STEEL BUILDING.
1. Air terminals with roof conductors bonded to building steel and looped to assure that a single wiring break will not fault the system. 2. Intermediate floor ground loops with connections to building mechanical equipment and piping to protect against side effect flashover to people or equipment. 3. Building steel, properly bonded together, used as down conductors. 4. Counterpoise conductor cross-connects ground electrodes with ground conductors located below minimum ground water level to assure adequate and uniform dissipation of the charge.

protection. The Empire State Building and the Washington Monument are two such systems, but they are slender in relation to their height and therefore a single rod on the uppermost part provides protection for the remainder of the structure. A single rod on a small chimney rising above a building may very well provide protection for the chimney but would in no way protect the remainder of the structure.

Specifications should require the installation of Underwriters Master Labeled lightning protection systems by a company so listed with the Underwriters Laboratories, Inc., Chicago. The Lightning Protection Institute of Chicago maintains a list of qualified installers of lightning protection.

Extreme care should be exercised in the design and specifications for structural steel and all high-rise buildings. Such buildings are not in themselves "lightning protected." Underwriters Laboratories states in their requirements that the structural steel framework of a building may be utilized as part of the lightning protection system but that this type of installation is "highly specialized and should not be undertaken by persons without full knowledge and training."

Lightning protection codes which may be referred to in specifying lightning protection systems are: "Installation Requirements for Master Labeled Lightning Protection System, UL96A," published by Underwriters Laboratories, Inc.; "Lightning Protection Code, 1965," NFPA No. 78, published by the National Fire Protection Assn., Boston; and in the case of buildings housing the manufacture, handling, or storage of ammunitions, explosives or flammable liquids and gases, reference may be made to the United States Army Ordinance Safety Manual.

Appendix:
Fire Protection
Reference Standards

Volume 10

Volume 11

Volume 12

Volume 13

National Model Building Codes

BOCA Basic Building Code
Building Officials and Code Administrators, International
1313 E. 60th Street
Chicago, Illinois

National Building Code
American Insurance Association
85 John Street
New York, N.Y. 10038

Southern Standard Building Code
Southern Building Code Congress
1116 Brown-Marx Building
Birmingham, Alabama

Uniform Building Code
International Conference of Building Officials
5360 Workman Mill Rd.
Whittier, California 90601

Index

About the Editor

Rolf Jensen is a leading authority on fire protection. He is
chairman of the Department of Fire Protection Engineering
at Illinois Institute of Technology, and president of Rolf
Jensen & Associates, Inc., fire protection and building code
consultants. The firm serves a number of large architectural
firms and other organizations. Previously, he directed the
Underwriters Laboratories engineering group testing fire
detection and extinguishing equipment.

Jensen has originated major developments in automatic
sprinkler protection for specialized industrial storage. In 1967
he headed a blue-ribbon committee investigating the McCormick
Place fire in Chicago and later designed the fire protection systems for the new building. He is now researching application
of systems analysis concepts to fire protection.